Introduction to Organic Chemistry I

11ᵗʰ Hour

Introduction to Organic Chemistry I

Seth Elsheimer

Department of Chemistry
University of Central Florida
Orlando, Florida

Blackwell
Science

©2000 by Blackwell Science, Inc.

Editorial Offices:
Commerce Place, 350 Main Street, Malden, Massachusetts 02148, USA
Osney Mead, Oxford OX2 0EL, England
25 John Street, London WC1N 2BL, England
23 Ainslie Place, Edinburgh EH3 6AJ, Scotland
54 University Street, Carlton, Victoria 3053, Australia
Other Editorial Offices:
Blackwell Wissenschafts-Verlag GmbH, Kurfürstendamm 57, 10707 Berlin, Germany
Blackwell Science KK, MG Kodenmacho Building, 7-10 Kodenmacho Nihombashi, Chuo-ku, Tokyo 104, Japan

Distributors:
USA
    Blackwell Science, Inc.
    Commerce Place
    350 Main Street
    Malden, Massachusetts 02148
    (Telephone orders: 800-215-1000 or 781-388-8250; fax orders: 781-388-8270)

Canada
    Login Brothers Book Company
    324 Saulteaux Crescent
    Winnipeg, Manitoba, R3J 3T2
    (Telephone orders: 204-224-4068)

Australia
    Blackwell Science Pty, Ltd.
    54 University Street
    Carlton, Victoria 3053
    (Telephone orders: 03-9347-0300; fax orders: 03-9349-3016)

Outside North America and Australia
    Blackwell Science, Ltd.
    c/o Marston Book Services, Ltd.
    P.O. Box 269
    Abingdon
    Oxon OX14 4YN
    England
    (Telephone orders: 44-01235-465500; fax orders: 44-01235-465555)

All rights reserved. No part of this book may be reproduced in any form or by any electronic or mechanical means, including information storage and retrieval systems, without permission in writing from the publisher, except by a reviewer who may quote brief passages in a review.

Acquisitions: Nancy Hill-Whilton
Development: Jill Connor
Production: Louis C. Bruno, Jr.
Manufacturing: Lisa Flanagan
Interior design by Colour Mark
Cover design by Madison Design
Typeset by Best-set Typesetter Ltd., Hong Kong
Printed and bound by Capital City Press

Printed in the United States of America
00 01 02 03 5 4 3 2 1

The Blackwell Science logo is a trade mark of Blackwell Science Ltd., registered at the United Kingdom Trade Marks Registry

Library of Congress Cataloging-in-Publication Data

Elsheimer, Seth Robert.
    Introduction to organic chemistry I / Seth Elsheimer.
        p.   cm.—(11th hour)
    ISBN 0-632-04417-9
    1. Chemistry, Organic Outlines, syllabi, etc. I. Title. II. Title:
Introduction to organic chemistry 1. III. Series: 11th hour (Malden, Mass.)
QD256.5.E38 2000
547—dc21   99-16541
                                                                                                                    CIP

CONTENTS

11th Hour Guide to Success	vii
Preface	viii

1. Structure, Bonding, and Molecular Properties — 1
1. Octet Rule and Lewis Structural Formulas — 1
2. Ionic, Covalent, and Polar Covalent Bonds — 4
3. Atomic Orbitals, Sigma and Pi Bonds — 5
4. Hybridization and Molecular Shapes — 7
5. Formal Charge — 9
6. Resonance — 12
7. Acids and Bases — 13
8. Bond Dipoles and Molecular Dipoles — 15

2. Alkanes: Introduction to Organic Structures and Isomers — 21
1. Structure and Nomenclature of Alkanes and Alkyl Groups — 21
2. Isomers and a Shorthand Notation — 23
3. Cycloalkanes, *cis* and *trans* Substituents — 26
4. Structures and Physical Properties of Alkanes — 28
5. Conformational Analysis of Alkanes and Cyclohexanes — 30
6. Reactions of Alkanes: Free Radical Halogenation and Combustion — 34
7. Classes of Organic Molecules and Functional Groups — 37

3. Alkenes: Electrophilic Addition Reactions — 43
1. Structure and Nomenclature, *cis/trans*, and *E/Z* — 43
2. Degree of Unsaturation — 46
3. Electrophilic Addition of HBr to Alkenes — 48
4. Hydration of Alkenes — 51
5. Halogenation and Halohydrin Formation — 53
6. Alkene Hydrogenation — 55
7. Alkene Cleavage — 57
8. Radical Additions of HBr to Alkenes — 60
9. Alkene Polymerization — 62
10. Preparation of Alkenes: Elimination Reactions — 64

4. Alkynes — 71
1. Structure, Nomenclature, and Preparation — 71
2. Some Electrophilic Additions — 73
3. Hydration of Alkynes: Keto-Enol Tautomerism — 75
4. Reduction of Alkynes to Alkenes and Alkanes — 78
5. Acidity of Alkynes and Alkylation of Acetylides — 79
6. Alkyne Cleavage Reactions — 82

5. Stereochemistry — 89
   1. Isomer Classification — 89
   2. Chirality — 91
   3. Classification and Number of Stereoisomers — 93
   4. Assigning Configuration Around Stereogenic Centers — 96
   5. Fischer Projections — 98
   6. Polarimetry and Associated Terminology — 101
   7. Properties of Stereoisomers and Resolution of Racemic Mixtures — 103

MIDTERM EXAM — 110

6. Alkyl Halides: Substitution and Elimination Mechanisms — 113
   1. Structure, Nomenclature, and Properties of Alkyl Halides — 113
   2. Preparation of Alkyl Halides — 115
   3. Nucleophilic Substitution Reactions: S_N1 and S_N2 — 117
   4. Elimination Reactions: E1 and E2 — 120
   5. Competing Reactions: Summary of S_N1, S_N2, E1, and E2 — 124
   6. Organometallic Reagents from Alkyl Halides — 125

7. Dienes and Conjugation — 133
   1. Conjugation and Conjugated Dienes — 133
   2. 1,2 and 1,4 Addition, Kinetic Versus Thermodynamic Control — 136
   3. The Diels-Alder Reaction — 139

8. Aromatic Compounds — 148
   1. Aromaticity, Benzene, and Resonance — 148
   2. Aromatic Nomenclature — 150
   3. General EAS Reaction and Some Specific Examples — 152
   4. Substituent Effects and EAS Reactions — 154
   5. Nucleophilic Aromatic Substitution — 157
   6. Other Reactions of Aromatic Compounds — 159

9. Spectroscopy — 167
   1. Mass Spectrometry — 167
   2. Ultraviolet Visible — 169
   3. Infrared Spectroscopy — 171
   4. Nuclear Magnetic Resonance and ^{1}H-NMR — 173
   5. Carbon-13 NMR — 176

FINAL EXAM — 183

11TH HOUR GUIDE TO SUCCESS

The 11th Hour Series is designed to be used when the textbook doesn't make sense, the course content is tough, or when you just want a better grade in the course. It can be used from the beginning to the end of the course for best results or when cramming for exams. Both professors teaching the course and students who have taken it have reviewed this material to make sure it does what *you* need it to do. The material flows so that the process keeps your mind actively learning. The idea is to cut through the fluff, get to what you need to know, and then help you understand it.

Essential Background. We tell you what information you already need to know to comprehend the topic. You can then review or apply the appropriate concepts to conquer the new material.

Key Points. We highlight the key points of each topic, phrasing them as questions to engage active learning. A brief explanation of the topic follows the points.

Topic Tests. We immediately follow each topic with a brief test so that the topic is reinforced. This helps you prepare for the real thing.

Answers. Answers come right after the tests; but, we take it a step farther (that reinforcement thing again), we explain the answers.

Clinical Correlation or Application. It helps immeasurably to understand academic topics when they are presented in a clinical situation or an everyday, real-world example. We provide one in every chapter.

Demonstration Problem. Some science topics involve a lot of problem solving. Where it's helpful, we demonstrate a typical problem with step-by-step explanation.

Chapter Test. For more reinforcement, there is a test at the end of every chapter that covers all of the topics. The questions are essay, multiple choice, short answer, and true/false to give you plenty of practice and a chance to reinforce the material the way you find easiest. Answers are provided after the test.

Check Your Performance. After the chapter test we provide a performance check to help you spot your weak areas. You will then know if there is something you should look at once more.

Sample Midterms and Final Exams. Practice makes perfect so we give you plenty of opportunity to practice acing those tests.

The Web. Whenever you see this symbol the author has put something on the web page that relates to that content. It could be a caution or a hint, an illustration or simply more explanation. You can access the appropriate page through *http://www.blackwellscience.com*. Then click on the title of this book.

The whole flow of this review guide is designed to keep you actively engaged in understanding the material. You'll get what you need fast, and you will reinforce it painlessly. Unfortunately, we can't take the exams for you!

PREFACE

This book was designed and written with you in mind. The goal is to quickly review the most important parts of a standard first-semester organic chemistry course. The format and coverage are intentionally brief. The author assumes you have already completed, or are currently enrolled in, a first semester organic chemistry course. This book contains relatively little of the usual pedagogical material and detailed explanations found in a standard organic chemistry textbook. Only the most gifted student is likely to master the concepts using this book as the only information source. Where possible, the theoretical bases for observable phenomena are mentioned; however, the brief review format unavoidably requires less explaining and more summarizing. Those conscientious students seeking a more detailed analysis or theoretical background are encouraged to consult one of the many excellent organic chemistry textbooks available. There are also web supplements referenced throughout this book. Many of these contain more detailed explanations or more examples. These can be accessed via http://www.blackwellscience.com.

There is some duplication in chapters and subjects in the Organic Chemistry I and II modules to accommodate the variation in course content found among different institutions. For that reason, the last two chapters in this book (Aromatic Compounds, and Spectroscopy) are dually listed as the first two chapters in the Organic Chemistry II module.

Sincere thanks go to Nancy Hill-Whilton, Executive Editor, and Jill Connor, Assistant Development Editor, for their frequent enthusiastic encouragement and guidance. Thanks also to the talented production staff at Blackwell Science, Inc. I gratefully acknowledge the following professors and students who reviewed the manuscript and made helpful suggestions: Patricia Heiden, Michigan Technological University; Kate Graham, College of St. Benedict/St. John's University; Sandra Lamb, University of California, Santa Barbara; Richard Jarocz, University of Wisconsin, Sheboygan; Christy Kenny, Ursinus College; Shawn Alderman, St. John's University; and Nader Mazloom, Michigan Technological University.

This book is dedicated to Janice Elsheimer—wife, author, and educator.

Seth Elsheimer, Ph.D.
Associate Professor
University of Central Florida

CHAPTER 1

Structure, Bonding, and Molecular Properties

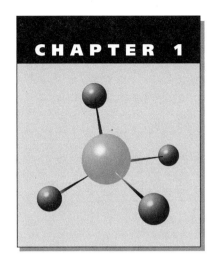

Historically, "organic chemistry" meant the chemistry of compounds that originated from living things. Such compounds were once believed to have a "vital force" that made them fundamentally different from those obtained from nonliving (inorganic) sources. As the vital force theory fell out of favor and was ultimately disproved, modern organic chemists observed that much of the chemistry of organic compounds resulted from the presence of carbon as the main element. Modern organic chemistry is the chemistry of carbon compounds.

ESSENTIAL BACKGROUND

- Atomic structure
- Valence electrons
- Electron configuration
- Periodic table
- Electronegativity
- Valence shell electron pair repulsion (VSEPR)

TOPIC 1: OCTET RULE AND LEWIS STRUCTURAL FORMULAS

KEY POINTS

✓ *What notations are used to represent molecules and bonds?*

✓ *How many bonds do atoms form when making compounds?*

There is a special stability associated with atoms that have a noble gas electron configuration. Atoms are normally bound within molecules in a way that leaves them with an electron configuration like that of a noble gas. Most atoms encountered in organic chemistry will become isoelectronic with helium or neon that have two and eight valence (outer shell) electrons, respectively. The popularity of the eight-electron arrangement is the basis for the **octet rule**. That is, one normally expects eight valence electrons around a bound atom (except for hydrogen, which normally has two). **Lewis dot structures** can be used to represent bonding. Dots represent bonded and nonbonded valence electrons, and the elemental symbol represents the atom's nucleus along with any core (nonvalence) electrons. The total number of valance electrons is determined from each atom's position in the periodic table with adjustments made for any charge on the species. A negative charge indicates an electron surplus, so one electron is added for each minus charge. A positive charge indicates an electron deficiency, so one electron is subtracted for each positive charge. Once the total number of valence electrons has been established, those electrons are distributed such that each atom except hydrogen has an octet (**Figure 1.1.a**). A common variation of the Lewis notation shows bonded pairs of electrons as lines and nonbonded pairs as dots (**Figure 1.1.b**).

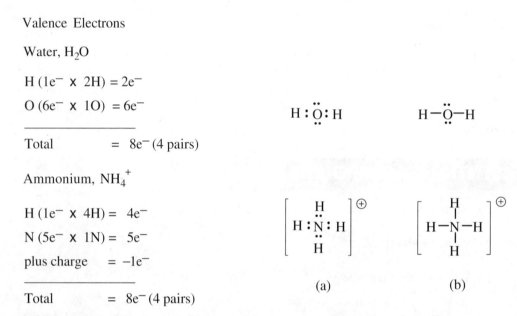

Figure 1.1 Valence Electron Counting and Lewis Representations of H_2O and NH_4^+.

Once you are more comfortable with the notation and its significance you may wish to omit nonbonded electron pairs from the picture. These are normally not shown explicitly unless you wish to emphasize their presence. An atom's position in the periodic table predicts the "normal" number of bonds required around a neutral (uncharged) atom to give it a noble gas configuration. A shared pair of electrons is considered as belonging to both atoms in this bookkeeping. Hydrogen forms one bond to become isoelectronic with helium, whereas the other elements listed bond to obtain an octet of valence electrons as shown in **Table 1.1**.

Table 1.1 Number of Bonds around Uncharged Atoms in a Molecule

Atom	Number of Bonds
H	1
C	4
N	3
O	2
X	1 (X = halogen = F, Cl, Br, I)

Chapter 1   Structure, Bonding, and Molecular Properties

Topic Test 1: Octet Rule and Lewis Structural Formulas

True/False

1. Most atoms form eight bonds.

2. A net negative charge on an ion indicates a surplus of electrons.

Multiple Choice

3. The usual number of bonds found around an uncharged carbon is
   a. 1
   b. 2
   c. 3
   d. 4
   e. None of the above

4. The total number of valance electrons in carbon dioxide is
   a. 8
   b. 10
   c. 12
   d. 16
   e. None of the above

5. Which of the following is the best structural formula for formaldehyde (CH_2O)?
   a. H—O=C̈—H

   b. 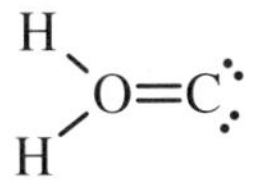 $\begin{array}{c}H\\ \diagdown\\ O=C\!:\\ \diagup\\ H\end{array}$

   c. :Ö=C⟨$\begin{array}{c}H\\ H\end{array}$

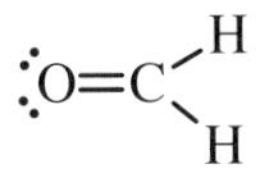

   d. 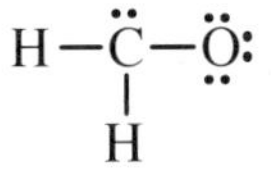 H—C̈—Ö:
      |
      H

   e. None of these is a correct structural formula.

Short Answer

6. What is the origin of the octet rule?

Topic Test 1: Answers

1. **False.** Most atoms form the number of bonds required to provide a stable electron configuration (often eight valence electrons). That is different from forming eight bonds.

2. **True.** Because electrons are negatively charged particles, a surplus of them would lead to a net negative charge.

3. **d.** An unbound carbon atom has four valence electrons. Forming four bonds will complete an octet.

4. **d.** The carbon provides four valence electrons and the two oxygens provide six each for a total of 16 valence electrons.

5. **c.** All other structures either have an incorrect number of valence electrons shown or have an incorrect number of electrons around one or more atoms.

6. The basis for the octet rule is the observation that many atoms bond to provide a filled valence shell having eight valence electrons (hydrogen is an exception).

TOPIC 2: IONIC, COVALENT, AND POLAR COVALENT BONDS

KEY POINTS

✓ *What is ionic bonding?*

✓ *What is covalent bonding?*

✓ *What is polar-covalent bonding?*

✓ *What types of bonding are most common in organic chemistry?*

Bonding between atoms that differ greatly in electronegativity is considered to result from electrostatic attraction between atoms or groups of atoms bearing unlike charges. These charged species are called ions, and this kind of bonding is called **ionic bonding**. Ionic compounds (salts) are often composed of elements from opposite sides of the periodic table. Usually these are elements from one of the two leftmost families of the periodic table (1A alkali metals or 2A alkaline earth metals) combined with the most electronegative families (6A chalcogens and 7A halogens). Examples include LiBr, NaCl, K_2O, MgF_2, and so on. Bonding between elements having little or no difference in electronegativity results from sharing valence electrons and is called **covalent**. Bonds between like atoms are excellent examples (H_2, Cl_2, or the central bond in ethane CH_3CH_3) but also included are the C—H bonds found in most organic molecules. Ionic and covalent actually are extremes of a bonding continuum. Atoms that differ in electronegativity by a moderate amount bond in a way that is somewhere between the extremes described above. Such bonds are called **polar covalent**. In these cases the electrons are shared, but the sharing is not equal. To visualize this, consider a cloud of electron density that surrounds the bonded atoms. The cloud is larger around the more electronegative atom. The rapidly moving electrons spend more time around the more electronegative atom. Examples of polar-covalent bonds are the O—H bonds of water or the C—Cl bonds in CCl_4. Most bonding in organic compounds is covalent and polar-covalent.

Topic Test 2: Ionic, Covalent, and Polar Covalent Bonds

True/False

1. Ionic compounds (salts) are often made of elements from the far right side of the periodic table combined with those from the far left of the periodic table.

2. The bonding in organic compounds is mostly covalent and polar-covalent.

Multiple Choice

3. The bonds in carbon dioxide molecules are
   a. ionic
   b. covalent
   c. polar covalent
   d. All of the above
   e. None of the above

4. The bonds that hold fluorine atoms together in F_2 are
   a. ionic
   b. covalent
   c. polar covalent
   d. All of the above
   e. None of the above

Short Answer

5. What is the major factor to consider when evaluating whether a bond is covalent, ionic, or polar covalent?

Topic Test 2: Answers

1. **True.** The far right and left sides of the periodic table are where the largest differences in electronegativity between combining elements exists.

2. **True.** Because carbon is near the middle of the electronegativity continuum, there are few species that differ from it significantly enough to give rise to ionic bonding.

3. **c.** The electronegativity of oxygen is sufficiently different from carbon to cause an unequal sharing of the electrons.

4. **b.** Although fluorine has a high electronegativity, the symmetrical F—F bond is equally shared between the two fluorine atoms.

5. The major factor is the electronegativity difference between the atoms involved. A large difference indicates ionic bonding, little or no difference indicates covalent bonding, and a moderate difference indicates polar covalent bonding.

TOPIC 3: ATOMIC ORBITALS, SIGMA AND PI BONDS

KEY POINTS

✓ *What are orbitals and how do they form bonds?*

✓ *What is the difference between sigma and pi bonds?*

✓ *How can one identify sigma or pi bonds in structural formulas?*

Orbitals are regions in space where electrons reside. Two types of atomic orbitals commonly encountered are s orbitals, which are spherical, and p orbitals, which are hourglass or dumbbell-shaped, as shown below in **Figure 1.2**. Any orbital can contain up to two electrons. Recall that there is one s orbital and three p orbitals in any energy shell (except the first, which has only the s orbital and no p orbitals) giving a total of four orbitals each of which can be doubly occupied for a possible total of eight electrons (octet rule).

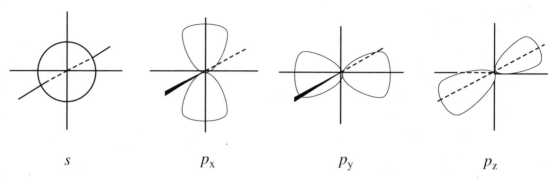

Figure 1.2 s and p Atomic Orbitals.

Covalent and polar-covalent bonds result from overlap of atomic orbitals. This overlap can be categorized as **sigma** (i.e., end-on, linear, cylindrical around the internuclear axis) or **pi** (side-on, parallel, overlap occurring above and below the internuclear axis). Sigma and pi bonds are illustrated in **Figure 1.3**. Generally, a single bond is sigma and pi bonds are found between atoms where multiple bonding occurs. The first bond between two atoms is normally a sigma bond and any additional bonds between these same atoms are pi bonds. For example, the carbon-carbon triple bond of ethyne (acetylene), H—C≡C—H is a sigma bond with two pi bonds superimposed over it. All four C—H bonds in methane, CH_4, are sigma.

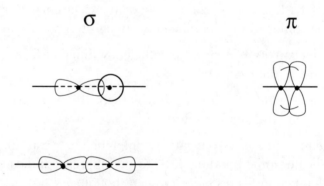

Figure 1.3 Sigma and Pi Bonds. •, nuclei; —, internuclear axes.

Topic Test 3: Atomic Orbitals, Sigma and Pi Bonds

True/False

1. Water contains no pi bonds.
2. Two spherical s orbitals can overlap to form a pi bond.

Multiple Choice

3. Which of the following is the best representation of a pi bond overlap?

6   Chapter 1   Structure, Bonding, and Molecular Properties

a.
b.
c.
d.

e. None of these represents a π bond.

4. Which of the following would you expect to contain pi bonds?
   a. Cyanide ion, :N≡C:⁻
   b. Carbon dioxide
   c. Ethene $H_2C=CH_2$
   d. All of the above
   e. None of the above

Short Answer

5. What are orbitals?

Topic Test 3: Answers

1. **True.** The bonds of water are single (sigma) bonds.
2. **False.** Pi bonds require a side-on or parallel overlap. Spherical s orbitals are the same on all sides and therefore cannot overlap sideways and are only involved in sigma bonds.
3. **a.** The other representations show an overlap that is end-on or cylindrical around the internuclear axis.
4. **d.** Each of the three species in a through c has multiple (double or triple) bonds between atoms. Normally the first bond is sigma and then any additional bonds are pi bonds.
5. Orbitals are regions in space where electrons reside.

TOPIC 4: HYBRIDIZATION AND MOLECULAR SHAPES

KEY POINTS

✓ *What are hybrid orbitals and how are they named?*

✓ *How is hybridization around a particular atom determined?*

✓ *How can VSEPR be used to predict bond angles and shapes?*

It is often convenient to imagine mixing (hybridizing) the atomic orbitals described above to provide a rational model for observed molecular formulas and structures. The number of **hybrid orbitals** resulting from this hybridization will be equal to the number of atomic orbitals hybridized (i.e., the total number of orbitals remains constant). These hybrid orbitals are named according to the number and kind of atomic orbitals that were hybridized to make them. For example, mixing the s and two of the three p orbitals will produce three new orbitals called sp² hybrid orbitals. Also, one unhybridized p orbital will be left over. The number of hybrid orbitals needed around a particular atom within a molecule is determined by the number of sigma bonds (do not count pi bonds!) and nonbonded electron pairs, as summarized in **Table 1.2**. Also listed in this table are shapes that result from maximum spatial separation of the hybrid orbitals. The terms listed under "Shape" describe molecular geometry (i.e., the spatial relationship of the atoms bound to the central atom). The electronic geometry is visualized by assuming that there is VSEPR.

Table 1.2 Hybridization and Molecular Shapes				
σ Bonds + Lone pairs	Hybridization	Approximate Angle	Shape	Example
4 + 0	sp³	109°	Tetrahedral	CH₄
3 + 1	sp³	109°	Trigonal pyramidal	NH₃
2 + 2	sp³	109°	Bent or angular	H₂O
3 + 0	sp²	120°	Trigonal planar	H₂C=CH₂
2 + 1	sp²	120°	Bent or angular	H₂C=NH
2 + 0	sp	180°	Linear	H-C≡C-H
1 + 1	sp	180°		H-C≡N:

Topic Test 4: Hybridization and Molecular Shapes

True/False

1. There are only three hybrid orbitals around an atom with sp³ hybridization.
2. The molecular shape of water is tetrahedral.

Multiple Choice

3. Which of the following contains an sp-hybridized carbon?

a. Propyne, H—C≡C—CH$_3$
   b. Ethanenitrile, N≡C—CH$_3$
   c. Carbon dioxide, O=C=O
   d. All of the above
   e. None of the above

4. The total sigma bonds plus nonbonded electron pairs around an sp^2-hybridized atom is
   a. 1
   b. 2
   c. 3
   d. 4
   e. None of the above

Short Answer

5. State the hybridization and molecular shape around the central carbon of CH$_2$O.

Topic Test 4: Answers

1. **False.** The name sp^3 hybridization indicates mixing of the s and all three p orbitals. Because the number of orbitals is constant (number of orbitals in equals number of orbitals out), there must be a total of four resulting sp^3 hybrid orbitals.

2. **False.** Water has two sigma bonds and two nonbonded electron pairs which leads to the prediction that it is sp^3-hybridized around oxygen. VSEPR theory tells us that the four resulting orbitals are directed to the corners of a tetrahedron; however, two of the orbitals do not connect to atoms. Although the electronic geometry is arguably tetrahedral with a bond angle near 109 degrees, the molecular shape is best described as bent or angular.

3. **d.** The carbon near the middle of each of a through c has two sigma bonds and no nonbonded electron pairs. This allows us to predict sp hybridization and a linear shape (bond angle 180 degrees). The leftmost carbon in a is also sp hybridized as is the nitrogen in b.

4. **c.** Recall that sp^2 hybridization results in three hybrid orbitals that can accommodate a total of three sigma bonds + nonbonded pairs.

5. After counting the total number of valence electrons for the entire species (12) and determining a structural formula consistent with the octet rule (H$_2$C=O), we find that the central carbon atom has three sigma bonds and no nonbonded electron pairs. The shape is therefore trigonal planar and the hybridization is sp^2.

TOPIC 5: FORMAL CHARGE

KEY POINTS

✓ *Why do some species have a charge whereas others do not?*

✓ *Where on a polyatomic ion is the charge represented on its structural formula?*

When the number of electrons around an atom differs from the number of electrons required for an uncharged isolated (unbound) atom, a **formal charge** exists. In such a case the formal charge *must* be shown to correctly specify the species. There is a significant difference between Na (soft gray metal that reacts explosively in moist air) and Na$^+$, which is commonly sprinkled on food. To determine the formal charge around a given atom, inspect the structural formula in which it is found to determine how many electrons are around the atom of interest. A non-bonded pair belongs exclusively to that atom, whereas a shared electron pair is regarded as belonging half to each atom for formal charge counting purposes. (Note this is different from the electron counting convention used for the octet rule.) The sum of the formal charges on a molecule or ion will equal the overall charge on that species. Formal charge can be calculated by subtracting the number of valence electrons in the bound atom from the number of valence electrons around the free atom.

$$\frac{\text{(no. valence e}^- \text{ around unbound atom)} - \text{(no. of nonbonding e}^- + \text{half the bonding e}^-\text{)}}{\text{formal charge}}$$

Some examples of structures where the sign, magnitude, and location of the formal charge are specified are shown in **Figure 1.4**.

Figure 1.4 Some Formal Charges Shown on Appropriate Atoms.

Topic Test 5: Formal Charge

True/False

1. The sum of the individual charges on the atoms in a molecule or ion will equal the net charge on the entire species.

2. Counting electrons for formal charge determination follows a different convention than counting electrons for determining compliance with the octet rule.

Multiple Choice

3. The formal charge on carbon in carbon monoxide, :C≡O:, is
   a. 0
   b. +1
   c. −1
   d. +4
   e. None of the above

4. The formal charge on oxygen in carbon monoxide, :C≡O:, is
   a. 0
   b. +1

c. −1
d. +4
e. None of the above

Short Answer

5. Among the unstable reactive intermediates discussed in organic chemistry are methyl carbocations (CH_3^+). Draw a structural formula and specify the location, sign, and magnitude of all nonzero formal charges.

6. Specify the location of any nonzero formal charges on the structural formulas below.

Topic Test 5: Answers

1. **True.**

2. **True.** When counting electrons for formal charge, only half of a shared electron pair (one electron) is considered to belong to each atom connected by that bond. Earlier we used the convention of counting for the octet rule in which the entire pair (two electrons) was considered as belonging to each atom connected by the bond. (For example, in CH_4 the carbon is thought to have an octet and the hydrogens are each valence satisfied at two electrons, which requires that each of the four C—H bonds be counted entirely for carbon and then again entirely for the hydrogen. All the atoms of methane have a formal charge of zero, however, which indicates that a different counting convention was used.)

3. **c.** There are five valence electrons shown around carbon for formal charge purposes. An uncharged carbon would require four, which gives a surplus of one electron.

4. **b.** There are five valence electrons showing around oxygen for formal charge purposes. An uncharged oxygen atom requires six electrons, which gives a deficiency of one electron.

5. Because there are only six total valence electrons, the octet rule cannot be satisfied. (In part that explains the high reactivity of this unstable species.)

6.

TOPIC 6: RESONANCE

KEY POINTS

✓ *What is the significance of resonance forms?*

✓ *Under what circumstances does one consider resonance?*

✓ *What are the conventions for writing resonance forms?*

Sometimes there is more than one correct Lewis structural representation for a molecule or ion. These will differ only by the positions of electrons but otherwise have the same connections among atoms and the same positions for the atoms involved. These structural formula representations are called **resonance forms**. No one of these resonance forms adequately describes the bonding, although each satisfies the usual rules for writing Lewis structures. The "real" species is best visualized as some combination of those resonance forms. The convention for indicating resonance forms is to place a double-headed arrow between them, as shown for the nitrite ion in **Figure 1.5**.

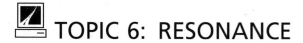

Figure 1.5 Resonance Forms of the Nitrite Ion.

Because the resonance forms are equivalent in this case, they contribute equally to the "real" structure. The bonds between nitrogen and oxygen are neither single nor double but rather are something in between. The negative charge is equally distributed over the two oxygens. The pi bond is not localized on one side of the ion as either resonance form alone would suggest but rather is delocalized over the three-atom array. Writing both forms and placing the double arrow between them tells the viewer that a nitrite ion is not really either of these inadequate representations but rather something a little like both. In general, resonance forms do not have to be equivalent. Each will have the same number and kind of atoms in the same locations but will differ only in the position of electrons. When drawing multiple resonance forms, do not move the atoms! Use care not to confuse the notation or concept of resonance with equilibrium (indicated by two arrows in opposite directions to show a reversible reaction) or an actual chemical transformation (indicated by a single headed arrow).

Topic Test 6: Resonance

True/False

1. Resonance is the way organic chemists represent several rapidly equilibrating species.
2. The structures O=N—O—H and H—O—N=O are resonance forms.

Multiple Choice

3. Which of the following is true about resonance forms?
   a. They are not always equivalent.
   b. They differ only in the position of electrons.

c. Together they represent a more stable species than any of them do individually.
d. All of the above
e. None of the above

4. What is the symbol used to represent resonance forms?
   a. An arrow $\longrightarrow$

   b. Two arrows in opposite directions, $\rightleftharpoons$

   c. A double headed arrow, $\longleftrightarrow$

   d. All of the above

   e. None of the above

Short Answer

5. Draw all reasonable resonance forms for the nitrate ion, NO_3^-. Place the correct symbol between resonance forms.

Topic Test 6: Answers

1. **False.** Resonance is not a process but rather a concept. Resonance forms are artificial or incomplete pictures that, if taken together, are used to represent a real structure that is lower in energy than any of the fictitious resonance forms. Rapidly equilibrating species are real and interconvert quickly due to a relatively low energy barrier between them.

2. **False.** These pictures represent the same thing drawn twice in the opposite directions. If two or more structures are resonance forms, they will have all the same atoms in the exact same locations and will differ only by the positions of electrons. Note that these two structural formulas have the hydrogen on different sides.

3. **d**

4. **c.** The single arrow in a is used to show a chemical transformation and the two arrows in b indicate equilibrium.

5.

TOPIC 7: ACIDS AND BASES

KEY POINTS

✓ *What are the Bronsted definitions for acid and base?*

✓ *What are conjugate pairs?*

✓ *What are conjugate acids and bases?*

✓ *What are the Lewis definitions of acid and base?*

Bronsted defined acids and bases in terms of a hydrogen ion (proton, H^+) being lost or gained. An **acid** is a proton donor and a **base** is a proton acceptor. The concept of conjugate pairs is a helpful tool in describing acid-base reactions. The species left behind after a Bronsted acid donates a proton is the **conjugate base** of that acid. Likewise, the base, once it has gained a proton in the acid-base exchange, becomes the **conjugate acid**. All these terms are illustrated in the generalized reactions below. Note that the species being transferred is formally H^+ (a hydrogen cation and not a neutral atom H or an anion H^-). The equations must be balanced both atomically and electrically.

$$\begin{array}{cccc} \text{Acid} & \text{Base} & \text{Conjugate Acid} & \text{Conjugate Base} \\ HY + Z & \rightarrow & HZ^+ & + & Y^- \\ H_2M^- + W^- & \rightarrow & HM^{2-} & + & HW \end{array}$$

The **Lewis** definition of acids and bases is focused on the electron pair that is supplied to a forming bond. The species that provides the electron pair is the **base** and the species accepting the pair is the **acid**. The Lewis definition is more general than the Bronsted definition. Some Lewis acid-base reactions do not even involve a proton.

$$\begin{array}{ccc} \text{Acid} & \text{Base} & \\ Mg^{+2} + 2\,H_2O & \longrightarrow & [Mg(OH_2)_2]^{+2} \\ BH_3 + CH_2=CH_2 & \longrightarrow & {}^+CH_2\text{-}CH_2\text{-}H_3B^- \end{array}$$

Topic Test 7: Acids and Bases

True/False

1. The conjugate base of HBr is hydroxide ion.
2. $AlCl_3$ can react as a Lewis acid but not as a Bronsted acid.

Multiple Choice

3. Which is the conjugate base of methanol, $HOCH_3$?
   a. $HOCH_3$
   b. $^+OCH_3$
   c. H_2OCH_3
   d. OCH_3
   e. None of the above

4. Which of the following is true about hydride ion, H^-?
   a. It is a Bronsted base.
   b. It is a Lewis base.
   c. It is the conjugate base of hydrogen gas, H_2.

d. All of the above.
e. None of the above.

Short Answer

5. Write a balanced chemical equation for the reactions of water with some strong acid, HA, and then do the same for the reaction of water with some strong base, B. In each reaction label the acid, base, conjugate acid, and conjugate base.

Topic Test 7: Answers

1. **False.** The conjugate base of HBr is bromide ion, Br⁻.
2. **True.** Aluminum trichloride has a vacant orbital that can accept a pair of electrons, making it a Lewis acid, but because there are no hydrogens in the formula for $AlCl_3$, it is not possible for it to react as a Bronsted acid (H^+ donor).
3. **e.** The conjugate base must differ by H^+ and will necessarily have one fewer hydrogen atom and one fewer plus charge. Answers b and d have the correct atoms but do not have the appropriate negative charge. The conjugate base of methanol is $^-OCH_3$.
4. **d.** All are true about hydride, which is both a Lewis and Bronsted base as well as being the conjugate base of H—H.
5. $HA + H_2O \rightarrow A^- + H_3O^+$
   acid   base   C.B.   C.A.

   $B + H_2O \rightarrow BH^+ + OH^-$
   base   acid   C.A.   C.B.

TOPIC 8: BOND DIPOLES AND MOLECULAR DIPOLES

KEY POINTS

✓ *How can one predict the magnitude and direction of bond dipoles?*

✓ *What notation represents polarities of bonds or molecules?*

Two common notations are used to represent bond polarity and molecular polarity. The lower-case Greek letter delta (δ) is used to indicate a partial (not full) charge. Alternately, the direction of the dipole can be shown with a vector pointed in the direction of the greater electron density or greater partial negative charge. Both these notations are illustrated below. Do not confuse partial charges and full formal charges.

$$\overset{\delta^+ \ \ \delta^-}{C\text{—}O} \qquad \overset{\longleftarrow\!+}{N\text{—}H}$$

These notations are also applied to entire molecules and not only just individual bonds. The polarity of a molecule can usually be deduced from vector addition of the individual bond polarities. For example, the polar bonds of water result in an overall molecular dipole due to the molecule's bent shape yet the two C=O bond dipoles of carbon dioxide cancel each other out, resulting in no net molecular dipole, as shown in **Figure 1.6**.

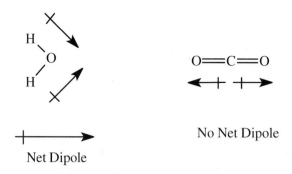

Figure 1.6 Bond Dipoles and Molecular Dipoles in Water and Carbon Dioxide.

Topic Test 8: Bond Dipoles and Molecular Dipoles

True/False

1. The bonds of CCl_4 are polar but the overall molecule has no net dipole.
2. Ammonia has no net molecular dipole.

Multiple Choice

3. Which of the following best represents the polarity of a C—F bond?
   a. δ− δ+
      C—F
   b. ←—+
      C—F
   c. ⁺C—F⁻
   d. All of the above
   e. None of the above

4. Which of the following has a molecular dipole?
   a. H_2S
   b. CH_2Cl_2
   c. FCH_3
   d. All of the above
   e. None of the above

Short Answer

5. Draw a structural formula for formaldehyde, CH_2O, and use the vector notation to show the individual bond polarities and the overall molecular dipole as needed.

Topic Test 8: Answers

1. **True.** Each C—Cl bond is polarized toward the more electronegative chlorine; however, the overall tetrahedral symmetry of the molecule leads to cancellation of the four bond dipoles and no net dipole for the molecule.

2. **False.** Ammonia, NH₃, has three polar-covalent bonds and a trigonal pyramidal shape. Vector addition for the three bond dipoles predicts a molecular dipole in the direction of the nitrogen.

3. **e.** Answer c shows full formal charges on the two atoms, which is incorrect for showing partial charges. Answers a and b use the correct notation for partial charges but show the negative and positive ends backward.

4. **d.** The bent geometry of H₂S does not allow the two H—S bond dipoles to cancel. The tetrahedral shape of CH₂Cl₂ requires that the two chlorine atoms are approximately 109 degrees apart and they do not cancel.

5. H
    \
     C=O
    /
    H

    +———→

DEMONSTRATION PROBLEM

Acetone has the formula C₃H₆O and is a three-carbon chain in which the center carbon is double bonded to an oxygen and the two outside carbons are connected to three hydrogens each. Draw all reasonable resonance forms for the conjugate base of acetone. Show all valence electrons, including both bonding and nonbonding electrons. Specify the location of all nonzero formal charges. Place the correct symbol between resonance forms.

Solution

Use the information given to draw a structural formula for acetone.

$$\begin{array}{c} \text{H} \quad :\!\text{O}\!: \quad \text{H} \\ | \quad\;\; || \quad\;\; | \\ \text{H}-\text{C}-\text{C}-\text{C}-\text{H} \\ | \quad\quad\quad\; | \\ \text{H} \quad\quad\;\; \text{H} \end{array}$$

The conjugate base of acetone will have one fewer H⁺ in its structure. Because acetone is symmetrical, it does not matter which proton is removed, but once that decision is made, any other resonance forms must have the same atoms in the same locations. The resonance forms should differ only by the position of electrons. Note that in this case the two resonance forms are not equivalent, so the "real" anion is not something half way between the two forms but rather is more like the form on the right that shows the negative charge on the more electronegative oxygen atom.

$$\begin{array}{c} \text{H} \quad :\!\text{O}\!: \\ | \quad\;\; || \\ \text{H}-\text{C}-\text{C}-\overset{..}{\text{C}}{}^{\ominus}-\text{H} \\ | \quad\quad\quad\; | \\ \text{H} \quad\quad\;\; \text{H} \end{array} \quad\longleftrightarrow\quad \begin{array}{c} \text{H} \quad :\!\overset{..}{\text{O}}\!:^{\ominus} \\ | \quad\;\; | \\ \text{H}-\text{C}-\text{C}=\text{C}-\text{H} \\ | \quad\quad\quad\; | \\ \text{H} \quad\quad\;\; \text{H} \end{array}$$

Chapter Test

True/False

1. The bonds of water are ionic.

2. A nitrogen surrounded by three sigma bonds and one nonbonded electron pair will have a formal charge of zero.

3. The symbol "↔" is used to indicate resonance forms.

4. Ethane, C_2H_6, contains only sigma bonds and no pi bonds.

Multiple Choice

5. The normal number of bonds found around an uncharged oxygen in an organic compound is
   a. 1
   b. 2
   c. 3
   d. 4
   e. 7

6. The total number of valence electrons in a carbonate ion, CO_3^{2-}, is
   a. 10
   b. 12
   c. 22
   d. 24
   e. None of the above

7. Which of the following molecular shapes can be found around atoms with sp^3 hybridization?
   a. Bent or angular
   b. Trigonal pyramidal
   c. Tetrahedral
   d. All of the above
   e. None of the above

8. Which hybridization is associated with bond angles of about 120 degrees?
   a. sp
   b. sp^2
   c. sp^3
   d. sp^4
   e. None of the above

Short Answer

9. Among the unstable reactive intermediates discussed in organic chemistry are methyl anions (CH_3^-) and methyl free radicals ($\cdot CH_3$). Draw structural formulas for each of these. Specify the location, sign, and magnitude of all nonzero formal charges.

10. Provide a structural formula for a compound with the formula CH_5N. Show all valence electrons, including any nonbonded electrons.

11. Describe the bond between carbon and nitrogen in CH_5N as sigma or pi.

12. Describe the bond between carbon and nitrogen in CH_5N as ionic, covalent, or polar-covalent.

13. What is the conjugate base of the dihydrogen phosphate ion, $H_2PO_4^-$?

14. State the hybridization and shape around each carbon in acetone (see Demonstration Problem).

15. Draw all reasonable resonance forms for the carbonate ion, CO_3^{2-}, and place the correct symbol between them. As usual, specify the location sign and magnitude of all nonzero formal charges.

16. In view of your answer to 15 above, predict the hybridization on carbon, the molecular shape, and whether or not there is a net molecular dipole for a carbonate ion.

17. Write a balanced chemical equation that shows the amide ion (NH_2^-) reacting with acetylene (H—C≡C—H) to yield ammonia and acetylide ion (H—C≡C:$^-$). Label each component of the reaction as acid, base, conjugate acid, or conjugate base.

18. The allyl cation has the formula $C_3H_7^+$ and its structure is a three-carbon chain with two hydrogens bound to each of the end carbons and one hydrogen bound to the central carbon. Draw all reasonable resonance forms for the allyl cation.

Chapter Test Answers

1. **False** 2. **True** 3. **False** 4. **True** 5. **b** 6. **d** 7. **d** 8. **b**

9. [structures of H–C̈–H with H below (two resonance forms shown)] 10. [structure of H–N̈–C–H with H's]

11. Sigma

12. Polar-covalent

13. HPO_4^{2-} hydrogen phosphate.

14. The center carbon is sp^2 hybridized and trigonal planar. The outside carbons are each sp^3 hybridized and tetrahedral.

15. [three resonance structures of carbonate ion shown]

16. sp^2, trigonal planar, no net molecular dipole.

17. H—C≡C—H + NH_2^- → H—C≡C:$^-$ + NH_3
     Acid   Base   Conj. Base   Conj. Acid

18. [two resonance structures of allyl cation shown]

Check Your Performance

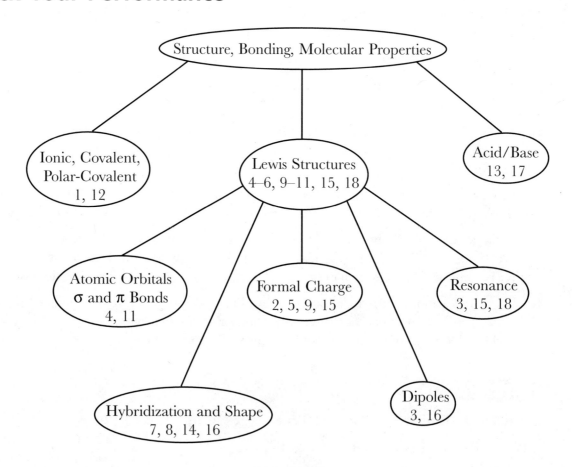

CHAPTER 2

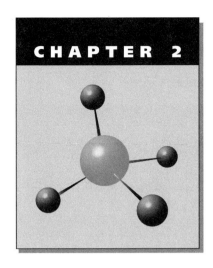

Alkanes: Introduction to Organic Structures and Isomers

Following your review and mastery of the important topics from Chapter 1, you are prepared to examine a simple class of organic compounds called **alkanes**. In this chapter we summarize the structure, nomenclature, properties, and reactions of alkanes and then set the stage for examining other classes of organic compounds.

ESSENTIAL BACKGROUND

- Lewis structural formulas (Chapter 1)
- Covalent bonds (Chapter 1)
- Sigma and pi bonds (Chapter 1)
- Hybridization and molecular shape (Chapter 1)
- Boiling point (BP), melting point (MP), density, van der Waals forces (General Chemistry)

TOPIC 1: STRUCTURE AND NOMENCLATURE OF ALKANES AND ALKYL GROUPS

KEY POINTS

- ✓ *What are hydrocarbons?*
- ✓ *What does "saturated" mean in relation to organic molecules?*
- ✓ *What are alkanes?*
- ✓ *How are unbranched alkanes named?*
- ✓ *What are alkyl groups and how are they named?*
- ✓ *How are branched alkanes named?*

Hydrocarbons contain only hydrogen and carbon. A maximum of 2n + 2 hydrogens can be found in a hydrocarbon with n carbons. Molecules that have the molecular formula C_nH_{2n+2} are called **alkanes** and are described as **saturated** because they contain the maximum number of hydrogens possible. Alkanes are composed of sp^3-hybridized tetrahedral carbon(s) that are single bonded to each other or to hydrogen atoms. Unbranched alkanes are named according to the number of carbons in the formula as summarized in **Table 2.1** as condensed structural

Table 2.1 Names and Condensed Structural Formulas for the First 12 Unbranched Alkanes

Methane	CH_4
Ethane	CH_3CH_3
Propane	$CH_3CH_2CH_3$
Butane	$CH_3CH_2CH_2CH_3$
Pentane	$CH_3CH_2CH_2CH_2CH_3$
Hexane	$CH_3CH_2CH_2CH_2CH_2CH_3$
Heptane	$CH_3CH_2CH_2CH_2CH_2CH_2CH_3$
Octane	$CH_3CH_2CH_2CH_2CH_2CH_2CH_2CH_3 = CH_3(CH_2)_6CH_3$
Nonane	$CH_3(CH_2)_7CH_3$
Decane	$CH_3(CH_2)_8CH_3$
Undecane	$CH_3(CH_2)_9CH_3$
Dodecane	$CH_3(CH_2)_{10}CH_3$

formulas. The unbranched alkanes with five or more carbons are named with a Greek prefix to indicate the number of carbons followed by the suffix "ane" to indicate it is an alkane.

It is often desirable to refer to a portion of a molecule or some fragment. Partial structures that resemble alkanes except that they have a hydrogen "missing" are called **alkyl groups** and generally have the formula C_nH_{2n+1}. Some alkyl groups are shown in **Figure 2.1** along with their common names. Note that they resemble their alkane analogues in structure and name except that they have a suffix "yl" in their name to indicate an "incomplete" structure.

$-CH_3$ $-CH_2CH_3$ $-CH_2CH_2CH_3$ $-CH-CH_3$ = $-CH(CH_3)_2$
                                                        $|$
                                                        CH_3

Methyl   Ethyl   Propyl   Isopropyl

$-CH_2CH_2CH_2CH_3$ $-CH_2CHCH_3$ = $-CH_2CH(CH_3)_2$ CH_3
                           $|$ $|$
                           CH_3 $-C-CH_3$ = $-C(CH_3)_3$
                                                            $|$
                                                            CH_3

Butyl   Isobutyl   tert-Butyl

Figure 2.1 Structures and Common Names for Some Simple Alkyl Groups.

Topic Test 1: Structure and Nomenclature of Alkanes and Alkyl Groups

True/False

1. Methanol, CH_3OH, is a hydrocarbon.
2. A fragment with the formula C_2H_5 is called ethane.

Multiple Choice

3. The correct name for an unbranched 10-carbon alkane is
   a. octane
   b. cyclodecane
   c. decane
   d. cyclohexane
   e. None of the above

4. How many hydrogen atoms will an alkane molecule with 22 carbons have?
   a. 10
   b. 22
   c. 44
   d. 46
   e. None of the above

Short Answer

5. Name the alkyl group that has the structure —$CH_2CH_2CH_3$
6. Name the compound with the structure $CH_3CH_2CH_2CH_3$

Topic Test 1: Answers

1. **False.** The presence of an oxygen in the formula excludes it from the hydrocarbons that by definition contain only hydrogen and carbon.
2. **False.** The correct name is eth*yl*. Alkyl groups (fragments) end with "yl." The "ane" suffix is reserved for alkanes. Eth*ane* has the formula C_2H_6.
3. **c.** The prefix "dec" indicates 10 carbons (as in decade or decimal) and the characteristic suffix "ane" identifies it as an alkane.
4. **d.** The number of hydrogens for an alkane with n hydrogens is 2n + 2. In this case, 2(22) + 2 = 46.
5. **Propyl.** The prefix "prop" indicates a three-carbon chain and the "yl" indicates that it is an alkyl group.
6. **Butane.** The prefix "but" indicates a four-carbon chain and the "ane" suffix specifies it as an alkane.

TOPIC 2: ISOMERS AND A SHORTHAND NOTATION

KEY POINTS

✓ *What are isomers?*

✓ *What are the International Union of Pure and Applied Chemistry (IUPAC) conventions for naming branched alkanes?*

✓ *How is the shorthand line-bond notation used to represent structural formulas?*

Molecules having the same formula but different structures are called **isomers**. For example, there are two molecules with the formula C_4H_{10} shown below. The first is the unbranched alkane butane and the second branched structure has the common or trivial name isobutane (i.e., an isomer of butane).

$$CH_3CH_2CH_2CH_3 \qquad CH_3\underset{\underset{\displaystyle CH_3}{|}}{C}HCH_3$$

Butane Isobutane

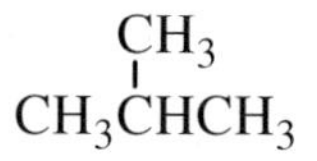

Similarly, there are three isomers with the formula C_5H_{12} as shown below along with their common names in parentheses. A more systematic nomenclature endorsed by IUPAC names alkanes as derivatives of the longest parent chain alkane in the structure. Any branches off the main parent are listed as alkyl groups. Numbers are used to indicate the positions of the branches, and numbering is from the end of the chain that gives the lowest number. The prefixes "di-," "tri-," and "tetra-" are used to indicate two, three, or four of the same alkyl substituents within a molecule, respectively. If two or more different-sized alkyl groups are part of the structure, they are listed in alphabetical order and each is given a locator number specifying where it is on the chain. The systematic IUPAC names for the C_5H_{12} isomers are shown below with the corresponding common names in parentheses.

$$CH_3CH_2CH_2CH_2CH_3 \qquad CH_3CHCH_2CH_3 \qquad CH_3-C-CH_3$$

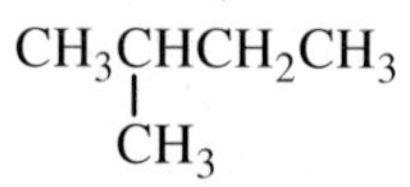

Pentane 2-Methylbutane 2,2-Dimethylpropane
 (Isopentane) (Neopentane)

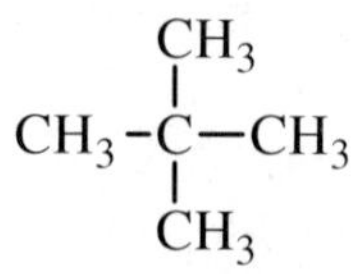

A shorthand notation exists for writing organic molecules. These are sometimes called line-bond structures or skeletal structures. The rules for writing these are summarized below.

1. Carbons are not explicitly shown but rather are represented by the ends, bends, and intersections of lines.

2. Hydrogens on carbon are also not explicitly shown. Sufficient hydrogens to give a total of four bonds around each carbon are assumed.

3. Other atoms (N, O, S, P, F, Cl, Br, I, etc.) are explicitly shown along with any hydrogens attached directly to them.

Applying these rules to the C_6H_{14} isomers gives the structures shown in **Table 2.2**.

Table 2.2 Structural Formulas and IUPAC Names for C_6H_{14} Isomers

Structure		IUPAC Name
(line structure) =	$CH_3(CH_2)_4CH_3$	Hexane
(line structure) =	$(CH_3)_2CHCH_2CH_2CH_3$	2-Methylpentane

(line structure) =	$CH_3CH_2CHCH_2CH_3$
	       $|$
	     CH_3
		3-Methylpentane

(line structure) =	$(CH_3)_3CCH_2CH_3$	2,2-Dimethylbutane
(line structure) =	$(CH_3)_2CHCH(CH_3)_2$	2,3-Dimethylbutane

Topic Test 2: Isomers and Shorthand Notation

True/False

1. There are three isomers with the formula C_3H_8.
2. The line-bond abbreviation for pentane is.

Multiple Choice

3. The IUPAC recommendation for naming branched alkanes specifies that
   a. the longest continuous chain is the parent.
   b. alkyl branches are listed in alphabetical order.
   c. the quantifiers di- and tri- indicate two or three of the same substituents, respectively.
   d. numbers are used to specify the location of substituents along the chain.
   e. All of the above

4. Which of the following is an isomer of octane?
   a. 2,3,4-Trimethyloctane
   b. 2,3-Dimethyloctane
   c. 2,2,3,3-Tetramethylbutane
   d. All of the above
   e. None of the above

Short Answer

5. Show a "line-bond" shorthand representation for $(CH_3CH_2)_2CHCH_2CH(CH_2CH_3)_2$.
6. Name the compound in problem 5 above.

Topic Test 2: Answers

1. **False.** Propane is the only compound with this formula. It can be drawn several ways, but all these represent a three-carbon chain with enough hydrogen atoms around the outside to give each carbon a total of four bonds.
2. **False.** This structure shows five connected lines that represent bonds (not atoms). The five lines connect six carbon atoms in a continuous chain, making this structure hexane.
3. **e**
4. **c.** Like octane, this compound has the formula C_8H_{18}. The alkanes in a and b have 11 and 10 carbons, respectively.
5.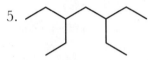
6. The longest continuous chain is seven carbons long (heptane derivative), and there are two branches, each of which contain two carbons (diethyl). These branches are bound to the parent chain on the third and fifth carbons. The name specifies all this information in one word: **3,5-diethylheptane**.

TOPIC 3: CYCLOALKANES, *CIS* AND *TRANS* SUBSTITUENTS

KEY POINTS

✓ *What are the IUPAC conventions for naming cycloalkanes?*
✓ *What do* cis *and* trans *mean and how are these terms used in naming cycloalkanes?*

As their name implies, cycloalkanes are alkanes that contain rings. Although they are considered a subset of alkanes, they will necessarily have two fewer hydrogens in the formula for each ring present and therefore a simple cycloalkane will have the formula C_nH_{2n}. They are named by inserting the prefix "cyclo" before the parent alkane name. Cycloalkanes often have alkyl groups appended, and, as usual, any alkyl substituents are indicated at the beginning of the name. When needed, numbers are included to specify the location of the alkyl groups. Some examples are show below.

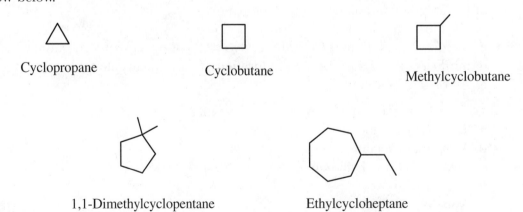

26 Chapter 2 Alkanes: Introduction to Organic Structures and Isomers

When a ring bears two or more substituents, it is often possible for them to have more than one spatial relationship. This is not immediately apparent unless one considers the molecule in three dimensions. Groups that point toward the same face of a ring are described as *cis* and those pointed toward opposite faces are called *trans*. These terms are added as prefixes to the name of a substituted cycloalkane as in the examples that follow.

cis-1,2-Dimethylcyclopropane

trans-1,2-Dimethylcyclopropane

trans-1-Ethyl-3-methylcyclobutane

cis-1-Ethyl-2-methylcyclobutane

Topic Test 3: Cycloalkanes, *cis* and *trans* Substituents

True/False

1. Cyclopentane has the formula C_5H_{12}.
2. There are two isomers of 1,1-dimethylcyclopropane and they are called *cis* and *trans*.

Multiple Choice

3. How many hydrogens are in propylcyclobutane?
   a. 7
   b. 12
   c. 14
   d. 16
   e. None of the above

4. The prefix *trans*- in the name of a cycloalkane indicates
   a. that two substituents point toward opposite faces of the ring.
   b. that two substituents point toward the same face of the ring.
   c. that two substituents are on the same ring carbon.
   d. that two substituents are on adjacent ring carbons.
   e. None of the above

Short Answer

5. How can one specify in a name that two ethyl groups on a cyclohexane are on neighboring carbons?

6. How can one indicate in the name of the compound in number 5 above whether the two ethyl groups are pointed toward the same face of the ring or opposite faces?

Topic Test 3: Answers

1. **False.** A cycloalkane with five carbons will contain only 10 hydrogens.

2. **False.** A 1,1-disubstituted cycloalkane will necessarily have its substituents pointed toward opposite faces of the ring because both are on the same carbon. There is no possibility of *cis/trans* isomers.

3. **c.** Because this cycloalkane has seven carbons (four on the ring and three on the propyl group), one can predict there will be 14 hydrogens from the 2n relationship. Alternatively, one can draw out the structure for propylcyclobutane and then count the hydrogens.

4. **a**

5. Numbers are used to indicate the location(s) of substituents. For neighboring carbons in a cycloalkane, the numbers used would be the lowest possible and sequential. One ethyl is assigned as attached to carbon 1 of the ring and the other would be, by definition, attached to carbon 2.

6. The prefixes *cis-* and *trans-* are used to indicate that ring substituents are pointing toward the same or opposite faces of the ring, respectively.

TOPIC 4: STRUCTURES AND PHYSICAL PROPERTIES OF ALKANES

KEY POINTS

✓ *How are carbons classified as primary, secondary, tertiary, and quaternary?*

✓ *What are some general physical and chemical properties of alkanes?*

✓ *How do BP, MP, and density relate to alkane structure?*

It is often desirable to classify a carbon according to how many other carbons are attached directly to it. The terms primary, secondary, tertiary, and quaternary are used to describe carbons that are attached directly to one, two, three, or four other carbons, respectively. Examples are shown in **Table 2.3**.

Table 2.3 Classification of Carbon Environments		
TYPE	**SYMBOL**	**EXAMPLES**
Methyl	0°	CH_3—H or CH_3—Br, etc.
Primary	1°	CH_3—CH_3 or the end carbons of $CH_3(CH_2)_{16}CH_3$
Secondary	2°	Any of the middle carbons in $CH_3(CH_2)_{16}CH_3$
Tertiary	3°	The central carbon in 2-methylpropane $HC(CH_3)_3$
Quaternary	4°	The central carbon in 2,2-dimethylpropane $C(CH_3)_4$

Alkanes are nonpolar and therefore have relatively low MPs and BPs compared with many other organic molecules. They are not water soluble to any appreciable amount, and they undergo relatively few reactions. In general, as the molecular weight of an alkane increases, so do its MP, BP, and density. Among isomers, the BP, MP, and density all decrease as the amount of branching increases. These trends are illustrated in the **Tables 2.4** and **2.5** and can be explained by the fact that decreasing the molecular surface area leads to decreased van der Waals attractions between the molecules.

Table 2.4 Physical Properties of C_1 through C_8 Unbranched Alkanes

Compound	BP (°C)	MP (°C)	Density (g/mL)
Methane	−162	−183	
Ethane	−89	−172	
Propane	−42	−188	
Butane	−0.5	−138	
Pentane	+36	−130	0.626
Hexane	+69	−95	0.659
Heptane	+98	−91	0.684
Octane	+126	−57	0.703

Table 2.5 BPs of C_5H_{12} Isomers

Compound	Structure	BP (°C)
Pentane	$CH_3CH_2CH_2CH_2CH_3$	36
2-Methylbutane (isopentane)	$(CH_3)_2CHCH_2CH_3$	28
2,2-Dimethylpropane (neopentane)	$C(CH_3)_4$	9.5

Topic Test 4: Structures and Physical Properties of Alkanes

True/False

1. Cycloheptane contains only secondary carbons.

2. Octane has a higher BP than hexane.

Multiple Choice

3. Alkanes are
   a. water soluble.
   b. polar.
   c. relatively unreactive compared with many other types of organic compounds.
   d. All of the above
   e. None of the above

4. Generally, as the amount of branching among alkane isomers increases, the
   a. BP increases.
   b. BP decreases.
   c. number of tertiary and quaternary carbons decreases.

d. number of primary carbons decreases.
e. None of the above

Short Answer

5. Draw a cycloalkane that contains at least one quaternary carbon.
6. How many tertiary carbons are found in the structure of 2-cyclopropyloctane?

Topic Test 4: Answers

1. **True.** The seven ring carbons of cycloheptane are each bound to two other carbons and are therefore classified as secondary.
2. **True.** The eight-carbon unbranched alkane octane will have greater surface area and therefore greater intermolecular van der Waals attraction than the six-carbon hexane.
3. **c**
4. **b.** Because there is a smaller surface area as the molecules are more branched (or more spherical), there is less intermolecular van der Waals attraction. Answers c and d are excluded by the observation that for each "branch" attached to a parent alkane chain, there will likely be one more primary carbon (the terminal methyl of the alkyl group) and one more tertiary or quaternary carbon at the site of attachment.
5. There are many possible correct answers here, including any 1,1-dialkylcycloalkane. Generally, they will be rings that have two alkyl groups bonded to the same ring carbon.
6. There are two tertiary carbons in 2-cyclopropyloctane. They are on either side of the bond that connects the cyclopropyl group to the octane chain.

TOPIC 5: CONFORMATIONAL ANALYSIS OF ALKANES AND CYCLOHEXANES

KEY POINTS

✓ *How do the terms "conformation" and "conformer" apply to molecular structure?*
✓ *What factors influence molecular conformation?*
✓ *What are Newman projections and how are they drawn?*
✓ *What is a dihedral angle and what do the terms "syn," "anti," and "gauche" mean?*
✓ *What conformers exist for cyclohexane and its derivatives?*
✓ *What do "axial" and "equatorial" mean relative to cyclohexane and its derivatives?*
✓ *What will be the most stable conformer of a particular substituted cyclohexane?*

Rotations around single bonds often allow a variety of possible spatial relationships among the atoms of molecules. The various **conformational isomers** or **conformers** will often have different relative stabilities. Generally, the stability of a given conformation will depend on how much strain exists in the structure. Molecular strain has three major components:

- **Torsional strain** resulting from the eclipsing of adjacent bonds;
- **Steric strain** resulting from simple crowding of atoms or groups into the same space;
- **Angle strain** resulting from deformation of bond angles away from the ideal.

Changes in **conformation** are often defined in terms of a **dihedral angle** as shown below. The tendency is for atoms and groups to separate themselves in a way that provides the least crowding. **Figure 2.2** summarizes some of the conformers of butane. These are defined according to the dihedral angle between the end methyl groups as they appear displayed in **Newman projections** that sight along the C(2)—C(3) bond. In these pictures, the "front" carbon (carbon 2) is represented by the intersection of the three lines in the center and the circle represents the "back" carbon (carbon 3).

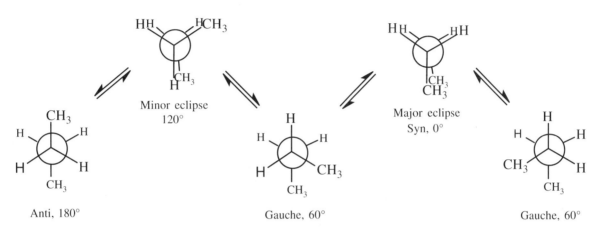

Figure 2.2 Butane Conformers.

Many cycloalkane rings are not flat but rather are puckered to accommodate the bond angles of sp³-hybridized carbons and to minimize the repulsions between nearby atoms. Conformations of six-membered rings illustrate these principles. The parent cyclohexane has several possible conformations illustrated in **Figure 2.3** along with the names suggested by their shapes.

Figure 2.3 Cyclohexane Conformers.

The "chair" conformers are the lowest in energy (most stable) and the "boat" is the highest (least stable). The hydrogens of cyclohexane are not all in the same environment for a given chair conformer. Half of them are straight up and down or parallel to the ring's axis, and these are called **axial**. The other half are out along the periphery of the ring and are approximately parallel to the ring's equator; these are called **equatorial**. When a cyclohexane ring undergoes a change from one chair conformer to the other (sometimes called "ring flip"), the axial and equatorial positions are reversed, as shown in **Figure 2.4**.

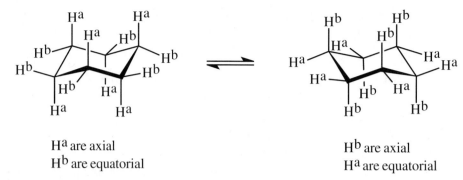

Figure 2.4 Axial and Equatorial Positions on Chair Cyclohexane.

Substituents on a cyclohexane ring tend to occupy equatorial positions where crowding is minimized. For example, the most stable conformer of methylcyclohexane is the chair that has the methyl in the equatorial position.

If more than one ring substituent is present, the favored conformer will generally be the one in which the largest substituent is equatorial as shown below for *trans*-1-isopropyl-3-methylcyclohexane.

Topic Test 5: Conformational Analysis of Alkanes and Cyclohexanes

True/False

1. *Cis*- and *trans*-1,2-dimethylcyclopropane are conformers.
2. The term "*gauche*" describes adjacent atoms or groups that have a 60-degree dihedral angle.

Multiple Choice

3. The most stable conformer of butane is known as
   a. *syn*
   b. *anti*
   c. *gauche*

Chapter 2 Alkanes: Introduction to Organic Structures and Isomers

    d. major eclipse
    e. minor eclipse
4. In general, substituents on chair cyclohexane rings prefer to be
    a. *syn*
    b. eclipsed
    c. axial
    d. equatorial
    e. None of the above

Short Answer

5. Draw a Newman projection for the most stable conformation of 2-methylbutane. Look along the C2–C3 axis.
6. Draw both chair conformers of *trans*-1,4-dimethylcyclohexane. Label each methyl group as axial or equatorial. Indicate which conformer is the most stable.

Topic Test 5: Answers

1. **False.** Conformers are conformational isomers and can interconvert through rotation around single bonds. The rigidity of the cyclopropane ring prevents *cis* methyl groups from becoming *trans* and vice versa. The two methyl groups in each of these isomers are constrained to be on the same or opposite sides of the ring, and that will not change unless bonds break.

2. **True.** *Gauche* describes the situation where the bonds of adjacent carbons are staggered (not eclipsed) and the atoms or groups under consideration appear 60 degrees apart in a Newman projection.

3. **b.** The *anti* conformer has a dihedral angle of 180 degrees and is lowest in energy due to the staggering of bonds and the large spatial separation of the terminal methyl groups

4. **d.** The equatorial positions provide more space and less steric crowding.

5. There are two methyl groups attached to carbon 2. The most stable conformer will place these at dihedral angles of 180 degrees (*anti*) and 60 degrees (*gauche*) to the methyl attached to carbon 3.

6.

Favored (most stable) conformer
Both methyl groups equatorial

Both methyl groups axial

TOPIC 6: REACTIONS OF ALKANES: FREE RADICAL HALOGENATION AND COMBUSTION

Key Points

✓ *How are free radical halogenations of alkanes carried out?*

✓ *What is a reaction mechanism?*

✓ *What are radicals?*

✓ *What are the mechanistic steps of a free radical alkane halogenation?*

✓ *Which positions on an alkane are the most reactive?*

✓ *What is the relative reactivity of Cl versus Br in radical halogenations?*

✓ *What will be the product(s) of a particular free radical halogenation reaction?*

✓ *What are the products of the complete combustion of hydrocarbons?*

There are relatively few reactions of alkanes, but among the most important of these are the **radical halogenations**. These can be initiated photochemically (hν) and work best when the halogen (X) is bromine or chlorine. The radical chlorination of methane is shown below.

$$CH_4 + Cl_2 \xrightarrow{h\nu} CH_3Cl + HCl$$

Radical halogenation reactions also often yield more highly halogenated products, and the reaction shown will likely also produce CH_2Cl_2, $CHCl_3$ and CCl_4. This can be minimized by using a large excess of alkane starting material.

A reaction **mechanism** is a description of what takes place along the path between reactant(s) and product(s), including which bonds break, which bonds form, the order of events, and the identification of any intermediates. Alkane halogenation reactions occur in several steps via intermediates, called radicals, that have an odd number of electrons. These are chain reactions that occur according to the steps shown in **Figure 2.5** for methane chlorination.

Initiation

1) $Cl\text{-}Cl \xrightarrow{h\nu} 2\,Cl^{\bullet}$

Propagation

2) $Cl^{\bullet} + CH_4 \longrightarrow CH_3^{\bullet} + HCl$

3) $CH_3^{\bullet} + Cl\text{-}Cl \longrightarrow CH_3Cl + Cl^{\bullet}$

Then 2) 3) 2) 3) ... etc., until ...

Termination

4) $CH_3^{\bullet} + Cl^{\bullet} \longrightarrow CH_3Cl$

Figure 2.5 The Reaction Mechanism for the Radical Chlorination of Methane.

For the radical halogenation of alkanes, chlorine is more reactive (less selective) and therefore more random than bromine. Complex product mixtures often result unless there is only one hydrogen environment as in methane shown above or cyclopentane below.

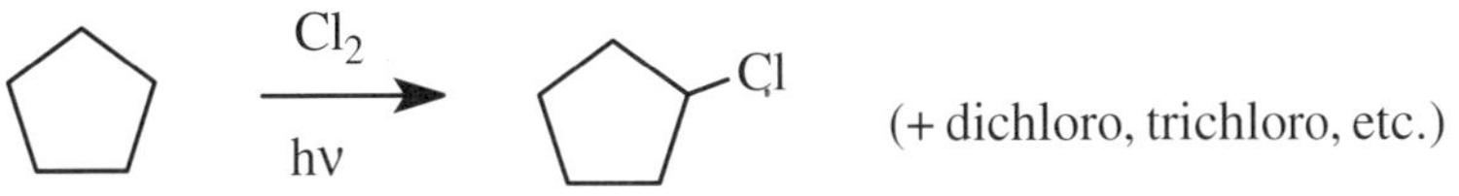

Bromine is less reactive (more selective), and the major product usually shows substitution in the preference order tertiary > secondary > primary > CH_3, as illustrated below. This selectivity reflects the relative stability of the required free radical intermediates.

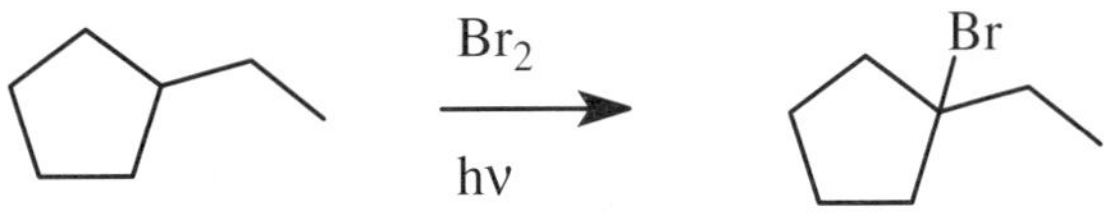

Another important reaction of alkanes is **combustion**. The complete combustion of any hydrocarbon is its reaction with oxygen to form carbon dioxide and water. The balanced equation for the complete combustion of propane is shown below.

$$CH_3CH_2CH_3 + 5\ O_2 \rightarrow 3\ CO_2 + 4\ H_2O$$

Topic Test 6: Reactions of Alkanes: Free Radical Halogenation and Combustion

True/False

1. Alkyl radicals such as methyl radical do not follow the octet rule.
2. Chlorine is more reactive and less selective than bromine in free-radical halogenation reactions of alkanes.

Multiple Choice

3. Which of the following could conveniently be made by a radical chlorination reaction of an alkane?

   a. 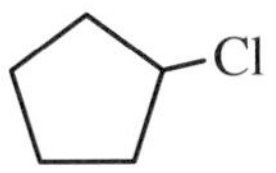
   b. 
   c. 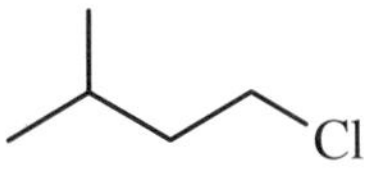
   d. 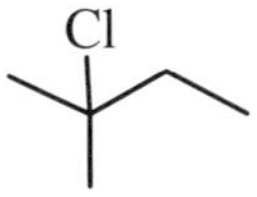
   e. All of the above

4. Which of the following could conveniently be made by a radical bromination reaction of an alkane?

a. [structure: (CH3)2CH-CH2-Br]

b. [structure: (CH3)2C(Br)-CH3... with Br on secondary/tertiary carbon]

c. [structure with Br on tertiary carbon]

d. [structure: CH2Br-CH(CH3)-CH3]

e. All of the above

Short Answer

5. Write a *balanced* chemical equation for the complete combustion of hexane.

6. What is a reaction mechanism?

Topic Test 6: Answers

1. **True.** A methyl radical has only seven valence electrons. Other alkyl radicals likewise do not have an octet of valence electrons around the radical carbon.

2. **True.** This explains the observation that mixtures of isomers result when some alkanes are chlorinated yet brominations preferentially take place at the most substituted carbons (tertiary > secondary > primary).

3. **a.** The chlorination of cyclopentane will take place cleanly because all 10 hydrogens of the reactant are equivalent and selectivity is not an issue. The chlorination of pentane in an attempt to make b or a chlorination of 2-methylbutane in an effort to make c or d would unavoidably lead to a mixture of products that differ by the position of the chlorine.

4. **c.** The bromine in this product is bound to a tertiary carbon, whereas the bromine in products a, b, and d are primary, secondary, and primary, respectively. The selectivity order is tertiary > secondary > primary for alkane bromination.

5. $2\ CH_3CH_2CH_2CH_2CH_2CH_3 + 19\ O_2 \rightarrow 12\ CO_2 + 14\ H_2O$

   These are easy to balance if one does the carbons first, the hydrogens next, and then the oxygen last. Fractional coefficients are acceptable so the answer can also be

   $$CH_3(CH_2)_4CH_3 + \tfrac{19}{2}\ O_2 \rightarrow 6\ CO_2 + 7\ H_2O$$

6. A reaction mechanism is a description of what takes place along the path between reactant(s) and product(s). This includes describing which bonds break, which bonds form, the chronological order of events, and the identity of any intermediates.

Chapter 2 Alkanes: Introduction to Organic Structures and Isomers

TOPIC 7: CLASSES OF ORGANIC MOLECULES AND FUNCTIONAL GROUPS

KEY POINTS

✓ *What is a functional group?*

✓ *What structures define some common functional classes of organic compounds?*

A collection of atoms in a particular bonding arrangement will have certain chemical characteristics that are common to all molecules containing that substructure. These are called **functional groups** and many are common in organic chemistry. Often the reactions we encounter for organic molecules can be generalized by showing the transformation of the functional group. Unless otherwise specified, the abbreviation "R" represents any alkyl portions of the molecule that do not undergo change. Some common functional classes of organic compounds are named along with generalized structures in **Table 2.6**.

Topic Test 7: Classes of Organic Molecules and Functional Groups

True/False

1. All alcohols and ethers contain oxygen.

2. A compound with the formula C_3H_9N is likely an amide.

Multiple Choice

3. Which of the following contain pi bonds?
   a. Alkenes
   b. Alkynes
   c. Aldehydes
   d. Ketones
   e. All of the above

4. The abbreviation "R" is often used in organic chemistry structural formulas to indicate
   a. a functional group.
   b. a reactive site on the molecule.
   c. a relatively unreactive alkyl group.
   d. All of the above
   e. None of the above

Short Answer

5. Name two functional groups that contain triple bonds.

6. Identify the functional groups present in the molecule shown below.

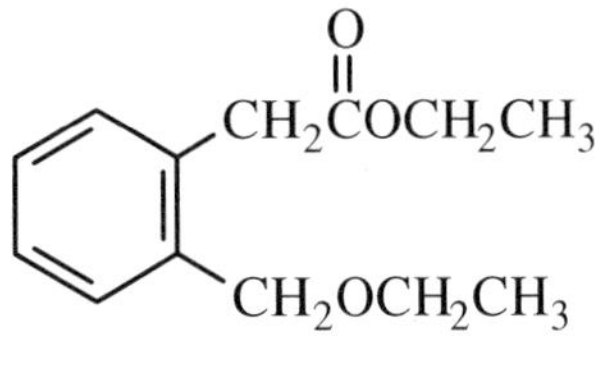

Table 2.6 Some Common Functional Groups

Name	Structure*		Example(s)
Alkane	R-H		$CH_3CH_2CH_3$, cyclopentyl-isopropyl
Alkene	$\diagdown C=C \diagup$		$(CH_3)_2C=CH_2$, propene
Alkyne	$-C\equiv C-$		$CH_3CH_2CH_2C\equiv CCH_3$
Alkyl halide	R-X	X = F, Cl, Br, I R = alkyl	$(CH_3)_2CHBr$ CH_3CH_2I
Aromatic	(benzene ring)		(phenyl)$-CH_3$
Alcohol	ROH	R = alkyl,	$CH_3CH_2CH_2OH$
Ether	ROR	R = alky or aryl	$CH_3CH_2OCH_2CH_3$
Aldehyde	$-\overset{O}{\underset{\|}{C}}-H$		$CH_3CH_2CH=O$ $H_2C=O$
Ketone	$R-\overset{O}{\underset{\|}{C}}-R$	R = alkyl or aryl	$CH_3\overset{O}{\underset{\|}{C}}CH_3$, cyclohexanone
Carboxylic acid	$-C\overset{O}{\underset{OH}{\diagdown}}$		CH_3CO_2H (phenyl)$-CO_2H$
(Carboxylic) ester	$-C\overset{O}{\underset{OR}{\diagdown}}$	R = alkyl or aryl	$CH_3CH_2\overset{O}{\underset{\|}{C}}OCH_3$
(Carboxylic) amide	$-C\overset{O}{\underset{N}{\diagdown}}$		$CH_3\overset{O}{\underset{\|}{C}}-NHCH_2CH_3$
(Carboxylic) acid halide or acyl halide	$-C\overset{O}{\underset{X}{\diagdown}}$	X = F, Cl, Br, I	$CH_3\overset{O}{\underset{\|}{C}}-Br$ (phenyl)$-COCl$
(Carboxylic) anhydride	$\overset{O}{\underset{\|}{C}}-O-\overset{O}{\underset{\|}{C}}$		$CH_3-\overset{O}{\underset{\|}{C}}-O-\overset{O}{\underset{\|}{C}}-CH_3$
Nitrile	$R-C\equiv N$	R = alkyl or aryl	$CH_3CH_2C\equiv N$ (phenyl)$-CN$
Amine	$-N\diagdown$		CH_3NH_2 $N(CH_2CH_3)_3$

*Unspecified bonds attach to H or C (alkyl or aryl).

Chapter 2 Alkanes: Introduction to Organic Structures and Isomers

Topic Test 7: Answers

1. **True.** Alcohols (ROH) and ethers (ROR′) each have oxygen atoms as part of their functional group, so any molecule that lacks oxygen cannot possibly belong to either of these classes.

2. **False.** The formula shows only C, H, and N. An amide would have to contain at least one oxygen because that atom is required for the amide functional group. The formula C_3H_9N is more likely an amine. (Students sometimes confuse the names amine and amide.)

3. **e.** Recall that pi bonds are present where multiple bonding (double or triple) is found between two atoms.

4. **c**

5. Two functional groups that contain triple bonds are alkynes and nitriles.

6.

Aromatic — (benzene ring) — $CH_2COCH_2CH_3$ (C=O) — Ester

$CH_2OCH_2CH_3$ — Ether

APPLICATION: OCTANE RATING FOR GASOLINE

Gasoline is a complex mixture of organic compounds including both branched and unbranched alkanes. The combustion of these compounds releases energy that powers automobiles. Branched alkanes burn more smoothly in automobile engines, and gasoline is rated (the "octane rating" displayed at the gasoline pump) according to how smoothly it will burn compared with a standard called isooctane (2,2,4-trimethylpentane, C_8H_{18}). A balanced chemical equation for the complete combustion of isooctane is

$$2\ C_8H_{18} + 25\ O_2 \rightarrow 16\ CO_2 + 18\ H_2O$$

DEMONSTRATION PROBLEM

Write a mechanism for the free radical bromination of 2-methylpropane.

Solution

The reactant 2-methylpropane contains three equivalent primary carbons and one tertiary carbon. Because the order of free radical stability is 3° > 2° > 1°, we predict that the tertiary radical will be favored over the primary radical and therefore the bromine substitution will occur at the central carbon.

Initiation
1. $Br\text{—}Br \xrightarrow{h\nu} 2\ Br^\bullet$

Propagation [Let R = $(CH_3)_3C$]
2. $Br^\bullet + R\text{—}H \rightarrow R^\bullet + HBr$
3. $R^\bullet + Br\text{—}Br \rightarrow RBr + Br^\bullet$

Then 2) 3) 2) 3) . . . until . . .

Termination

$R^\bullet + Br^\bullet \rightarrow RBr$

Other chain-terminating steps are possible, including $2\ R^\bullet \rightarrow R\text{—}R$

The net transformation is $Br_2 + (CH_3)_3CH \rightarrow (CH_3)_3CBr + HBr$

Chapter Test

True/False

1. Hexane and cyclohexane are isomers.
2. Secondary radicals are more stable than tertiary.
3. The BP of octane is greater than that of 2,2,3,3-tetramethylbutane.
4. Cyclohexane is water soluble.

Multiple Choice

5. Which of the following are hydrocarbons?
   a. Aldehydes
   b. Ketones
   c. Ethers
   d. Alkyl halides
   e. None of the above

6. Which of the following is produced by complete combustion of alkanes?
   a. Alcohols
   b. Alkyl halides
   c. Alkenes
   d. All of the above
   e. None of the above

7. The *syn* dihedral angle is
   a. 180 degrees
   b. 120 degrees
   c. 90 degrees
   d. 60 degrees
   e. 0 degrees

8. In the most stable conformer of *cis*-1,3-dimethylcyclohexane, the methyl groups are
   a. *trans*
   b. both axial
   c. both equatorial
   d. on the same carbon
   e. None of the above

Short Answer

9. Draw all the monochlorination products that result from the free radical chlorination of 2,5-dimethylhexane.

10. Name the following compound.

$$CH_3CHCH_2CH_2CHCH_2CH_3$$
$$\quad\;|\qquad\qquad\;\;|$$
$$CH_3CH_2\qquad CH_3$$

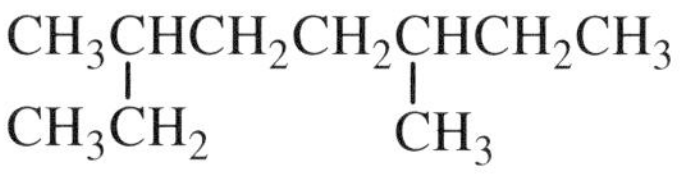

11. Name the following compound.

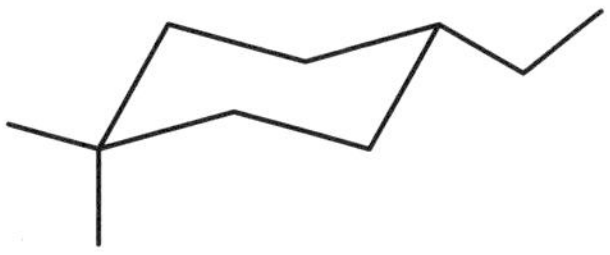

12. Use the line-bond shorthand notation to represent 1,1,4,4-tetraethylcycloheptane.

13. A common substructure in organic molecules is the carbon-oxygen double bond (C=O), which is called a carbonyl. Many of the functional groups mentioned in this chapter contain carbonyls. Name them.

14. Indicate the functional groups present in the following compound.

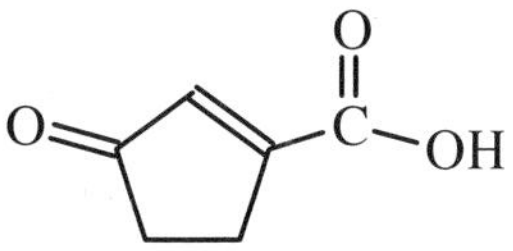

Essay

15. How does the relative stability of radicals explain the observed selectivity for free radical bromination reactions of alkanes. In other words, what is the mechanistic explanation for the observation that bromine usually ends up on the most substituted carbon.

Chapter Test Answers

1. **F** 2. **F** 3. **T** 4. **F** 5. **e** 6. **e** 7. **e** 8. **c**

9.

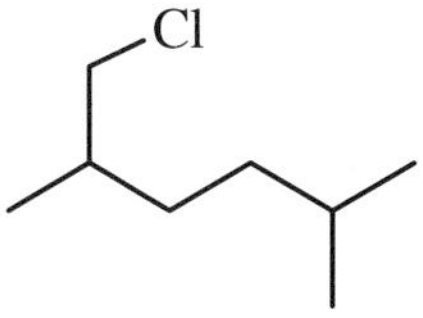

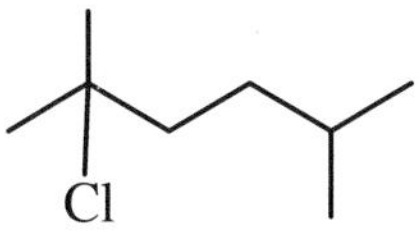

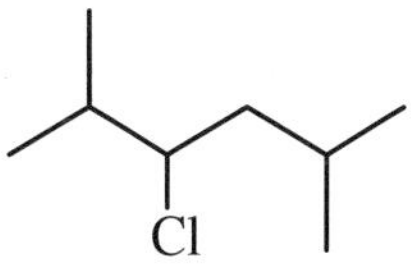

10. 3,6-Dimethyloctane

11. 4-Ethyl-1,1-dimethylcyclohexane

12.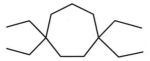

13. Aldehyde, ketone, carboxylic acid, ester, amide, acid halide, anhydride

14. Ketone, alkene, carboxylic acid

15. The first propagation step of the free radical chain mechanism involves abstraction of a hydrogen atom from the alkane by a bromine radical. The ease with which this abstraction takes place will depend on the stability of the alkyl radical produced in that step. If there is more than one hydrogen type, the easiest one to abstract is the one that leads to the most stable alkyl radical. Radical stability increases with greater substitution around carbon (i.e., 3° > 2° > 1° > 0°) methyl. After the alkyl radical is formed, the next propagation step places a bromine atom on the radical site. In other words, the location of the radical in step 2 dictates which product is ultimately formed in step 3. That is, bromine substitution is favored in an order that reflects the relative stability of the corresponding alkyl radicals.

Check Your Performance

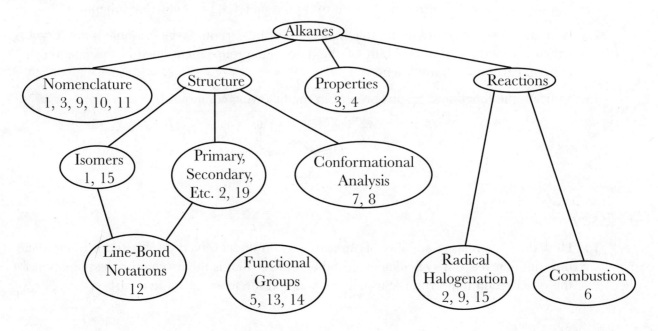

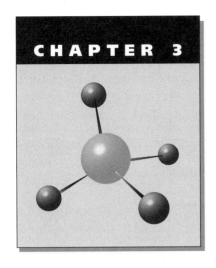

CHAPTER 3

Alkenes: Electrophilic Addition Reactions

Alkenes are everywhere and many have biological importance. The simplest alkene, ethene, is a plant hormone that regulates fruit ripening, flower maturation, and seed germination. Many of the pheromones that insects use to communicate alarm or find mates are alkenes, as are some of the flavors and fragrances found in plants. In this chapter we survey some of the chemistry of carbon-carbon double bonds. Much of this chapter sets the stage for understanding future topics in Organic I and II.

ESSENTIAL BACKGROUND

- Drawing structural formulas (Chapter 1)
- Alkane nomenclature (Chapter 2)
- Functional groups: alkyne, alcohol, aldehyde, ketone, carboxylic acid (Chapter 2)

TOPIC 1: STRUCTURE AND NOMENCLATURE, *cis/trans*, AND *E/Z*

KEY POINTS

✓ *What are the hybridization and shape of the carbons in an alkene?*
✓ *How are alkenes named?*
✓ *What do* cis *and* trans *mean as applied to alkenes?*
✓ *What are the Cahn-Ingold-Prelog (CIP) rules for prioritization?*
✓ *How are the CIP rules used to name alkene isomers as E or Z?*

Alkenes are hydrocarbons containing a carbon-carbon double bond. Simple acyclic alkenes have the general formula C_nH_{2n} and contain two sp²-hybridized carbons connected by a sigma bond with a pi bond superimposed as shown below in the structure of ethene.

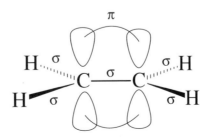

Unbranched alkenes names have the form "n-____ene," where the blank part is the parent prefix name indicating the number of carbons in the chain (prop, pent, hept, etc.) and "n" is a number identifying the position in the chain where the first double-bonded carbon is encountered. Any alkyl group branches are named at the beginning with numbers indicating where they are attached. Numbering begins at the end of the chain nearest to the alkene linkage and gives the lowest number possible to the alkene. Multiple alkene linkages are specified as dienes, trienes, and so on as shown in **Table 3.1**.

Table 3.1 Structures and Names of Some Alkenes

Structure	Name
$CH_2=CH_2$	Ethene
$CH_2=CHCH_3$	Propene
$CH_2=CHCH_2CH_3$	1-Butene
$CH_3CH=CHCH_3$	2-Butene
$CH_2=CHCH_2CH_2CH_3$	1-Pentene
$CH_3CH_2CH=CHCH_3$	2-Pentene
$CH_2=CHCH_2CH(CH_3)_2$	4-Methyl-1-pentene
$CH_2=CH-CH=CH_2$	1,3-Butadiene
$(CH_3)_2C=CHCH_2CH=C(CH_3)_2$	2,6-Dimethyl-2,5-heptadiene

Some alkenes require an additional prefix to specify the orientation of groups around the double bond. For example, 2-butene has two forms that differ by whether the methyl groups are on the same side or opposite sides of the double bond. These are called *cis* and *trans*, respectively.

cis-2-Butene   *trans*-2-Butene

A more general way to specify *cis/trans* relationships is based on the priorities of the groups attached to each of the alkene carbons. The groups are prioritized using rules proposaled by Cahn, Ingold, and Prelog (sometimes called the CIP rules):

1. Consider the atoms attached directly to the alkene carbons. These are prioritized by atomic number (i.e., I > Br > Cl > F > O > N > C > H). On each of the two alkene carbons there will be a higher and a lower priority substituent. If the two high-priority groups are *cis* to one another, the prefix designator is *Z* (German: *zusammen* = same) and if the high priority groups are *trans* the designator is *E* (German: *entgegen* = opposite).

2. If the atoms bound to the alkene carbons are the same, then go out to the second or third bonds away from the pi bond to prioritize the groups.

3. A double or triple bond between atoms is equivalent to two or three single bonds between atoms.

The following examples illustrate these rules.

E   *Z*   *E*

44   Chapter 3   Alkenes: Electrophilic Addition Reactions

There are two unsaturated hydrocarbon groups that have recognized common names. Use of these is illustrated below along with the International Union of Pure and Applied Chemistry alternative.

—CH=CH$_2$
Vinyl
(Ethenyl)

—CH$_2$CH=CH$_2$
Allyl
(2-Propenyl)

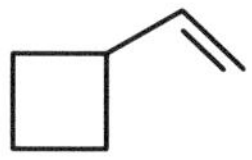

Vinylcyclobutane

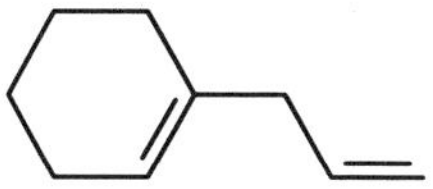

1-Allylcyclohexene

Topic Test 1: Structure and Nomenclature, cis/trans, and E/Z

True/False

1. All six atoms of ethene are in the same plane.
2. The CIP priorities are based on size.

Multiple Choice

3. Which of the following has the highest CIP priority?
   a. —CH(CH$_3$)$_2$
   b. —CH$_2$CH$_2$CH$_3$
   c. —CH$_2$CH=CH$_2$
   d. —CH$_2$CH$_2$CH$_2$CH$_3$
   e. —BH$_2$

4. Which of the following has cis and trans isomers?
   a. 2-Methyl-2-butene
   b. 1-Heptene
   c. Cyclobutene
   d. None of the above
   e. All of the above

Short Answer

5. Name the following compound.

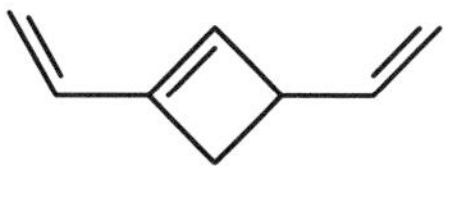

6. Draw an unambiguous structural formula for (Z)-3-cyclopropyl-2-pentene.

Topic Test 1: Answers

1. **True.** The sp² hybridization of the alkene carbons gives them trigonal planar geometry and the pi bond between them requires parallel overlap of the p orbitals, thus forcing all six atoms to be coplanar.

2. **False.** Size is not important. The priorities are based on the atomic number of the attached atom. For example a t-butyl group has a lower priority than a fluorine atom even though the former is much larger.

3. **a.** The center carbon of this isopropyl group is attached to a hydrogen and two other carbons (secondary). The hydrocarbon groups b, c, and d are all primary as drawn. Boron has a lower atomic number than carbon, so e is ruled out.

4. **d.** Compounds a and b have two of the same ligands (methyls or hydrogens) on one of their alkene carbons so unavoidably one of the ligands will be *cis* to whatever is on the other alkene carbon and the other will be *trans*. Another way to see this is to imagine switching the two methyl groups on C₂ of 2-methyl-2 butene. Is the resulting compound the same or different from what you started with? Cyclobutene cannot exist as *trans* due to the geometry around the sp²-hybridized carbons and the small inside angles of the four-membered ring. Try to draw the fictitious *trans*-cyclobutene and you will see.

5. This is a four-membered ring cycloalkene with two vinyl groups on carbons 1 and 3 of the ring (carbons 1 and 2 are the alkene carbons by definition). The name therefore is 1,3-divinylcyclobutene or 1,3-diethenylcyclobutene.

6.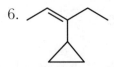

TOPIC 2: DEGREE OF UNSATURATION

KEY POINTS

✓ What does "degree of unsaturation" mean?

✓ How does one determine the degree of unsaturation?

✓ What does the degree of unsaturation reveal about a compound?

Saturated molecules contain the maximum number of hydrogens possible. For hydrocarbons this is C_nH_{2n+2}. For each ring present there will be two fewer hydrogens. Likewise, there will be two fewer hydrogens for each pi bond present. By inspection of a molecular formula, it is often possible to predict the total number of rings and/or pi bonds in the structure. Extending this logic past hydrocarbons to include compounds containing oxygen, nitrogen, and halogen is relatively easy. Because an oxygen atom normally forms two bonds, its presence in a structure does not change the maximum number of hydrogens, and no adjustment is needed for the $2n + 2$ expectation. Nitrogen normally forms three bonds, and its presence requires one additional hydrogen atom for each nitrogen atom in the formula. Halogens, like hydrogen, are normally monovalent and, in effect, take the place of hydrogen in a saturated formula. The number of hydrogens expected for a saturated molecule containing any combination of C, N, O, and X is given by

Number of hydrogen expected for $C_nN_aO_bX_x = 2n + 2 + a - x$

For every two hydrogens fewer than this found in the formula there is one **degree of unsaturation**, that is, one pi bond or one ring. Two examples of molecular formulas, degree of unsaturation, and structures are shown below.

Name	Formula	Degree of Unsaturation	Structure
Caffeine	$C_8H_{10}N_4O_2$	6	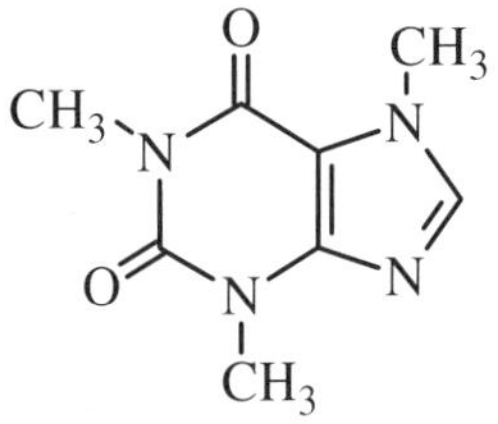
"DDT" (Insecticide)	$C_{14}H_9Cl_5$	8	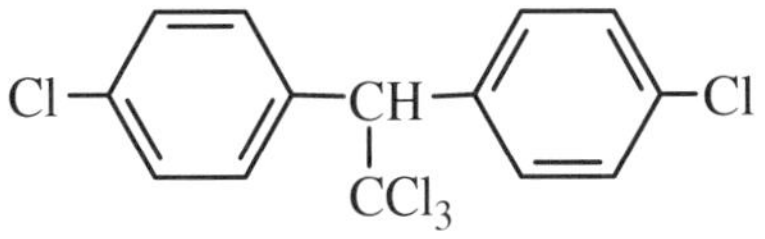

Topic Test 2: Degree of Unsaturation

True/False

1. 1,4-Cycloheptadiene has two only degrees of unsaturation.

2. A carbonyl will give a formula one degree of unsaturation.

Multiple Choice

3. Which formula represents a saturated compound?
   a. $C_5H_{10}BrF$
   b. $C_5H_{12}O_4$
   c. $C_5H_{13}N$
   d. All of the above
   e. None of the above

4. What is the minimum degree of unsaturation possible for an alkyne?
   a. 1
   b. 2
   c. 3
   d. 0
   e. None of the above

Short Answer

5. A bicyclic (two-ring) compound has the formula C_8H_{10}. How many pi bonds are in the structure?

6. How many isomers have four carbons and one degree of unsaturation?

Topic Test 2: Answers

1. **False.** The name 1,4-cycloheptadiene indicates presence of a ring and two pi bonds that total three degrees of unsaturation.
2. **True.** The double bond of the carbonyl is a sigma and a pi bond, the latter of which gives one unsaturation.
3. **d**
4. **b.** An alkyne has a triple bond between two carbons. One bond is a sigma and the other two are pi bonds. Each pi bond accounts for a degree of unsaturation.
5. **Two.** The formula corresponds to 4 degrees of unsaturation. Because two of these are known to be rings, the other two must be pi bonds.
6. There are five C_4H_8 isomers: cyclobutane, methylcyclopropane, 1-butene, 2-methylpropene, *cis*-2-butene, and *trans*-2-butene.

TOPIC 3: ELECTROPHILIC ADDITION OF HBr TO ALKENES

KEY POINTS

✓ *What are the conventions for writing reaction mechanisms?*
✓ *What is the mechanism for HBr addition to alkenes?*
✓ *What do the terms "electrophile" and "nucleophile" mean?*
✓ *What is Markovnikov's rule and what is the basis for this observation?*
✓ *What is a carbocation?*
✓ *What is the relative order of carbocation stability?*
✓ *Under what circumstances will a carbocation rearrange?*
✓ *How can one predict the product of HBr addition to a given alkene?*

When an alkene reacts with HBr, an alkyl bromide is produced. The hydrogen and bromine usually end up on the former alkene carbons. Unsymmetrical alkenes yield products in which the hydrogen goes on the less substituted carbon, that is, the alkene carbon that had the most hydrogens. This predictive tool is called **Markovnikov's rule**, which is explained by the **mechanism**. The conventions for writing and discussing the mechanism require that curved arrows indicate the direction of electron flow. A species with a negative or partial negative charge (called a **nucleophile**) is shown attacking a species with a positive or partial positive charge (called an **electrophile**). The structures of any intermediates along the path between reactant(s) and product(s) must be shown. The mechanism for electrophilic addition of HBr to propene is shown below.

$$CH_3CH=CH_2 \longrightarrow CH_3-\overset{\oplus}{C}H-CH_3 \longrightarrow CH_3-\underset{Br}{\overset{|}{C}H}-CH_3$$

$$H-Br \qquad Br^{\ominus}$$

$$(\text{not } CH_3CH_2\overset{\oplus}{C}H_2)$$

The key intermediate is the **carbocation**, resulting from the first step. The bromide ion (nucleophile) attaches to the carbocation (electrophile). The location of the charge on the carbocation dictates what product will be observed. The relative stability of carbocations is 3° > 2° > 1° > methyl. The first step in the mechanism forms the most stable carbocation and therefore determines which isomeric product is formed. When predicting products of HBr addition to alkenes, imagine the most stable carbocation intermediate and then the result of bromide attack on that electrophile. Under some circumstances, the carbocation intermediate rearranges to a more stable carbocation before it has the chance to be trapped by the attacking bromide. These rearrangements involve the migration of a hydride or an alkyl group adjacent to the carbocation center. In general, this should be considered when the rearrangement involves a 1,2 shift of a hydride or alkyl group and the shift results in a more stable carbocation. Some examples follow.

Topic Test 3: Electrophilic Addition of HBr to Alkenes

True/False

1. Most electrophiles have a negative charge.
2. A primary carbocation is more stable than a tertiary carbocation.

Multiple Choice

3. The first step in the mechanism for electrophilic addition of HBr to an alkene is
   a. attachment of the bromide to the most substituted carbon.
   b. a carbocation rearrangement.
   c. protonation of an alkene carbon to yield a carbocation.
   d. All of the above
   e. None of the above

4. When writing a mechanism, one should
   a. use curved arrows to indicate the direction of electron flow.
   b. show the structure(s) of any intermediate(s).
   c. specify the location of all nonzero formal charges.
   d. All of the above
   e. None of the above

Short Answer

5. Complete the following chemical equation with unambiguous structural formula(s) for the missing organic product(s).

6. Complete the following chemical equation with an unambiguous structural formula for the missing organic reactant.

Topic Test 3: Answers

1. **False.** As the name electrophile ("electron-lover") implies, an electrophile seeks out electron density and is itself electron deficient. Electrophiles usually bear positive or partial positive charges.

2. **False.** Carbocation stability increases in the order $CH_3^+ < 1° < 2° < 3°$.

3. **c.** Responses a and b are ruled out because the bromide cannot attach to the molecule or a carbocation cannot rearrange until after the carbocation is formed.

4. **d**

5.

6.

Chapter 3 Alkenes: Electrophilic Addition Reactions

TOPIC 4: HYDRATION OF ALKENES

Key Points

✓ *What is alkene hydration?*

✓ *What are three common methods for alkene hydration?*

✓ *How do the hydration methods differ in outcome?*

✓ *What is the mechanism for acid-catalyzed alkene hydration?*

✓ *What product(s) is obtained from a given hydration reaction?*

The addition of water across the double bond of an alkene is called **hydration**. The simplest hydration strategy is the acid-catalyzed reaction, which obeys Markovnikov's rule and proceeds via the mechanism illustrated below for propene.

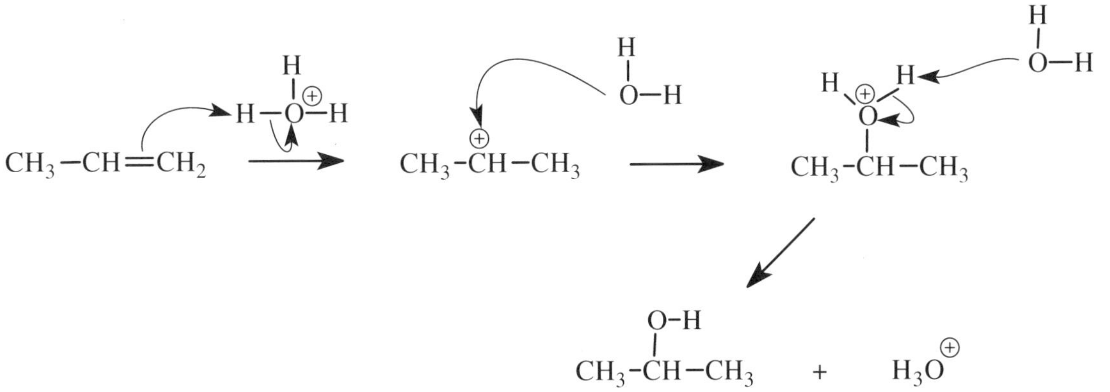

The acid-catalyzed hydration is subject to carbocation rearrangements; however, this problem can be overcome by using a two-step process often called **oxymercuration** (also sometimes called oxymercuration-demercuration).

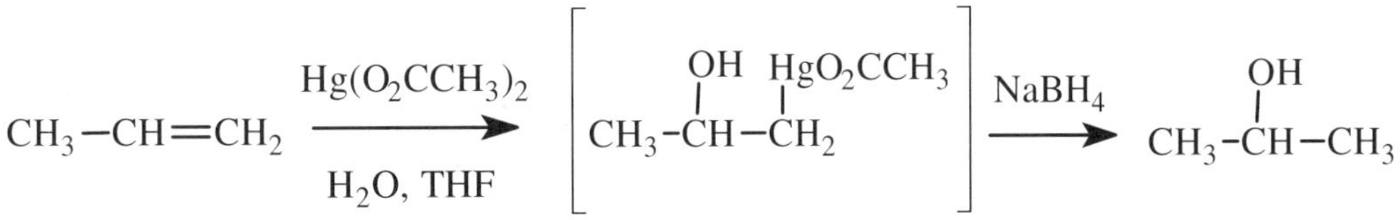

A third strategy for hydration of alkenes is the two-step process via **hydroboration**. This method gives a non-Markovnikov final product. In the first step, the H—B bond adds across the alkene with *syn* orientation (i.e., the H and the B go to the same face of the planar alkene). The boron attaches to the least substituted alkene carbon. The second step replaces the boron by an OH in the same position. The net transformation is an alkene to a non-Markovnikov alcohol. The *cis/trans* orientation of the final product can be predicted by considering the partial mechanism as illustrated for 1-methylcyclopentene below.

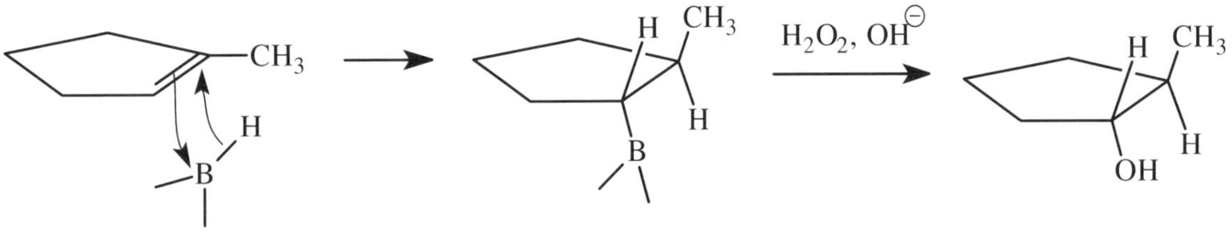

Topic Test 4: Hydration of Alkenes

True/False

1. Alcohols can be produced from alkene hydration reactions.

2. Aqueous mercury (II) acetate followed by sodium borohydride are reagents to hydrate an alkene with Markovnikov orientation and reduced risk of carbocation rearrangement.

Multiple Choice

3. Which conditions below would convert 1-butene into $CH_3CH_2CH_2CH_2OH$?
   a. H_2O/H_3O^+
   b. $Hg(O_2CCH_3)_2$ followed by $NaBH_4$
   c. BH_3 followed by H_2O_2/OH^-
   d. All of the above
   e. None of the above

4. Which is true about the mechanism for acid-catalyzed alkene hydration?
   a. The catalyst is regenerated in the last step.
   b. Water is the nucleophile.
   c. There is a carbocation intermediate.
   d. All of the above
   e. None of the above

Short Answer

5. Complete the following chemical equation with unambiguous structural formula(s) for the missing organic product(s).

   (CH₃)₂C=CHCH₃ →[Hg(O₂CCH₃)₂ / H₂O, THF] →[NaBH₄]

6. Complete the following chemical equation with an unambiguous structural formula for the missing organic reactant.

   ___ →[BH₃] →[H₂O₂, OH⁻] cyclopentyl-CH₂OH

Topic Test: Answers

1. **True.** Adding water across the pi bond produces an alcohol.
2. **True.** The conditions Hg(O₂CCH₃)₂ followed by NaBH₄ produce a Markovnikov alcohol without the risk of carbocation rearrangements.
3. **c.** These are the only conditions we have seen that lead to a non-Markovnikov alcohol.
4. **d**
5. 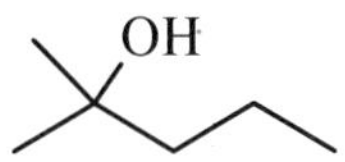 [structure: 2-methyl-2-pentanol with OH]
6. 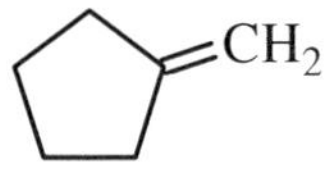 [structure: methylenecyclopentane]

TOPIC 5: HALOGENATION AND HALOHYDRIN FORMATION

KEY POINTS

✓ What are 1,2-dihalides (vicinal dihalides) and how are they prepared?

✓ What are halohydrins and how are they prepared?

✓ What is the mechanism for these reactions?

The molecular halogens Br₂ and Cl₂ react with alkenes to yield products called 1,2-dihalides, or **vicinal dihalides**, in which the halogen atoms are on adjacent (formerly alkene) carbons. The mechanism is similar to some of the electrophilic additions to alkenes above except that a key intermediate, the cyclic **halonium ion**, must be attacked by the halide nucleophile on its opposite face, resulting in exclusive *anti*-addition of X₂. If the alkene reactant is cyclic, as illustrated below, this mechanism requires that the halogens end up *trans* in the product.

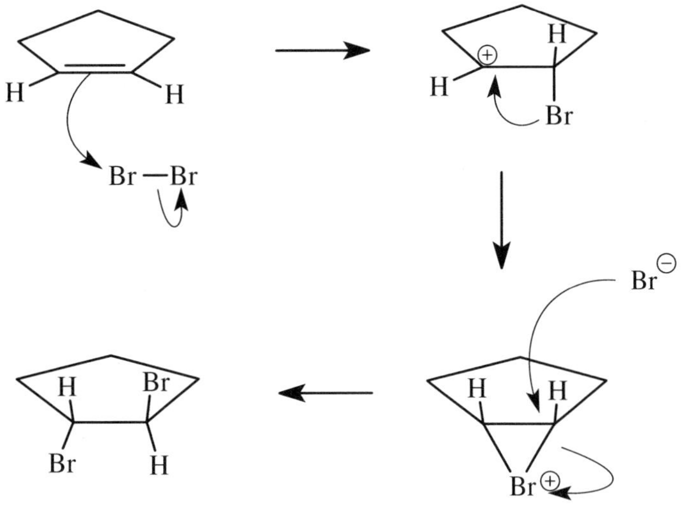

When the reaction is carried out in water, the nucleophile water rather than a halide ion can trap the bromonium intermediate. The resulting product is a halo-alcohol called a **halohydrin**. The OH is on the more substituted of the former alkene carbons as shown below.

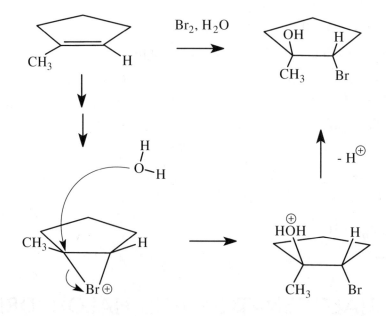

Topic Test 5: Halogenation and Halohydrin Formation

True/False

1. The product of an alkene halogenation reaction will have two halogen atoms on the same carbon.

2. The mechanisms for halogenation and halohydrin formation proceed via the same cyclic halonium ion intermediate.

Multiple Choice

3. When cycloalkene is treated with Cl_2/H_2O, the likely product
   a. has two chlorine atoms that are *cis*.
   b. has two chlorine atoms that are *trans*.
   c. is a chlorohydrin with the Cl and OH *cis*.
   d. is a chlorohydrin with the Cl and OH *trans*.
   e. None of the above

4. The nucleophile that attacks the cyclic halonium ion during a halohydrin-forming reaction is
   a. a halide ion.
   b. a hydroxide ion.
   c. a halogen molecule, X_2.
   d. water.
   e. None of the above

Short Answer

5. Complete the following chemical equation with an unambiguous structural formula(s) for the missing organic product(s).

54 Chapter 3 Alkenes: Electrophilic Addition Reactions

[structure: methylenecyclopentane + Br₂, H₂O →]

6. Complete the following chemical equation with an unambiguous structural formula(s) for the missing organic reactant.

[_____ + Br₂/CCl₄ → 1,1-dimethyl-substituted cycloheptane with trans Br, H, H, Br substituents]

Topic Test 5: Answers

1. **False.** The two halogen atoms in the product are vicinal, that is, they are on adjacent carbons that were formerly connected by a double bond.

2. **True.** The early steps in the mechanisms for each of these reactions are the same and they both involve attack of a nucleophile on a cyclic halonium ion.

3. **d.** The mechanism requires that the cyclic chloronium intermediate is attacked from the unhindered side, resulting in net anti-addition of the OH and Cl and therefore *trans* ring substituents.

4. **d.** The cyclic halonium intermediate is trapped by a water molecule during the step that forms the C—O bond. Subsequent proton abstraction from the protonated alcohol yields the observed halohydrin.

5. [1-(bromomethyl)cyclopentan-1-ol structure]

6. [cycloheptene with gem-dimethyl group]

TOPIC 6: ALKENE HYDROGENATION

KEY POINTS

✓ *What do the terms "oxidation" and "reduction" mean in organic chemistry?*

✓ *Under what conditions are alkenes hydrogenated (reduced) to alkanes?*

Oxidation and reduction are sometimes more subtle or difficult to identify in organic reactions than they are in inorganic chemistry. If Fe^{2+} reacts to form Fe^{3+}, this is clearly a loss of electrons or, perhaps even easier, an increase in oxidation number. Organic compounds have less obvious oxidation states. The following generalizations aid in identifying an organic reaction as oxidation or reduction.

Oxidation: A decrease in the electron density around carbon; replacing a less electronegative atom or group on carbon with a more electronegative atom or group; a decease in the number of hydrogens or an increase in the number of oxygens.

Reduction: An increase in the electron density around carbon; replacing a more electronegative atom or group attached to carbon by a less electronegative atom of group; an increase in the number of hydrogens or a decrease in the number of oxygens.

The hydrogenation of alkenes to alkanes is a reduction. This addition reaction places hydrogen atoms on the two former alkene carbons and produces an alkane of similar structure. The reaction is normally carried out with hydrogen gas in the presence of a transition metal catalyst like Pd/C (Pd deposited on charcoal) or PtO_2 (Adams' catalyst). Normally, the added hydrogen atoms attach to the same face of the alkene as shown below.

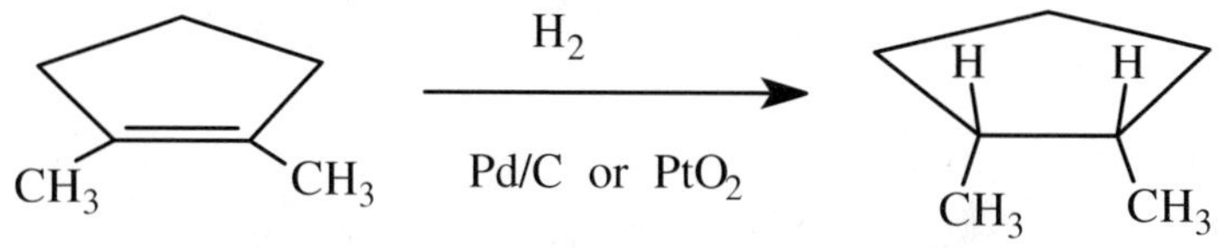

Topic Test 6: Alkene Hydrogenation

True/False

1. Alkene hydrogenation is an example of an organic reduction reaction.
2. An alkene hydrogenation reaction normally results in placement of two hydrogens on the same carbon of an alkene.

Multiple Choice

3. Which of the following is an oxidation?
   a. The combustion of methane to CO_2 and H_2O
   b. The conversion of methane to CH_2F_2
   c. The conversion of cyclohexane to benzene (C_6H_6)
   d. All of the above
   e. None of the above

4. An alkene hydrogenation reaction is normally catalyzed by
   a. acid
   b. hydrogen
   c. metal
   d. solvent
   e. None of the above

Short Answer

5. Complete the following chemical equation with an unambiguous structural formula(s) for the missing organic product(s).

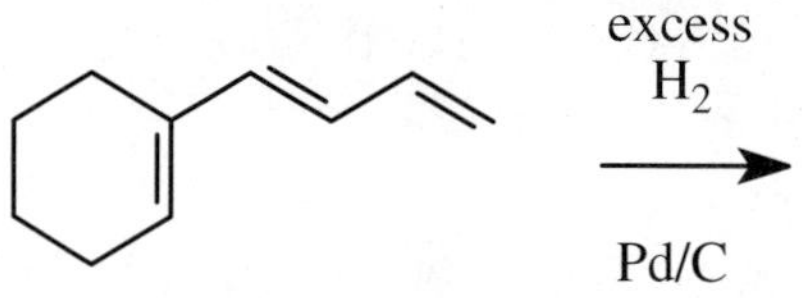

6. Reduction of an organic molecule can usually be identified by (complete each phrase with "increase" or "decrease").

Chapter 3 Alkenes: Electrophilic Addition Reactions

a. a(n) _____ in the electron density around carbon.
   b. a(n) _____ in the number of oxygens.
   c. a(n) _____ in the number of hydrogens.

Topic Test 6: Answers

1. **True.** The increase in the number of hydrogen atoms going from an alkene to an alkane qualifies this as a reduction. Because hydrogen is arguably more electronegative than nothing at all, one could also consider this a decrease of electron density around the alkene carbons.

2. **False.** The reaction places two hydrogens on *adjacent* carbons and normally on the same face (*syn* addition).

3. **d.** The combustion in response a is a loss of hydrogen, a gain in oxygen, and a decrease in electron density around carbon ($CH_4 \rightarrow CO_2$). Response b shows replacement of two hydrogens with fluorine (i.e., decrease in hydrogen or, better yet, replacement of hydrogen with a more electronegative atom resulting in decreased electron density around carbon). Response c is a dehydrogenation of C_6H_{12} to C_6H_6 and therefore qualifies as an oxidation.

4. **c.** The most common hydrogenation catalysts are based on transition metals (Pd, Pt, etc.). Do not be confused by response b. Although hydrogen is an essential part of most hydrogenation reactions, it is a reactant and not a catalyst.

5.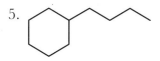

6. a, increase; b, decrease; c, increase.

TOPIC 7: ALKENE CLEAVAGE

KEY POINTS

✓ What are three strategies for cleaving alkenes?

✓ What reagent sequence is used for each alkene cleavage strategy?

✓ What are the intermediate products for each alkene cleavage strategy?

✓ What are the final products from each alkene cleavage strategy?

When treated with acidic potassium permanganate ($KMnO_4$), an alkene cleaves. Where a C=C bond formerly existed, there are now two **carbonyls**. Disubstituted alkene carbons yield ketones, whereas monosubstituted alkene carbons lead to aldehydes that further oxidize under these conditions to carboxylic acids. Terminal alkene carbons (=CH_2) are cleaved to formaldehyde, which is further oxidized to carbonic acid that in turn decomposes to carbon dioxide and water. These results are generalized below where R ≠ H and the final products are circled.

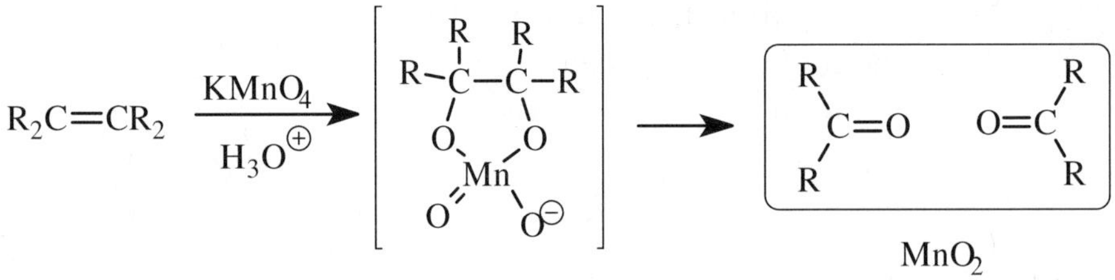

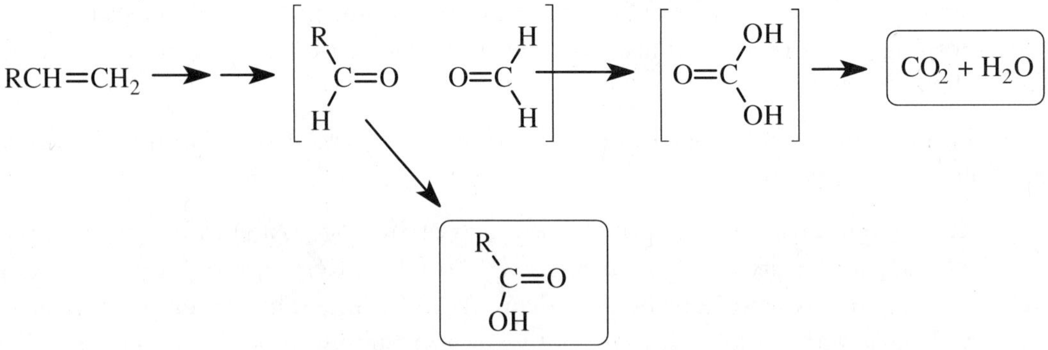

$R_2C=CR_2$ $\xrightarrow[H_3O^{\oplus}]{KMnO_4}$ $\left[\begin{array}{c} RR \\ R-C-C-R \\ OO \\ \diagdown Mn \diagup \\ O^{} O^{\ominus} \end{array}\right]$ ⟶ $\boxed{\begin{array}{cc} R & R \\ \diagdown C=O \quad O=C \diagup \\ R & R \end{array}}$

MnO_2

$RCH=CH_2$ ⟶ ⟶ $\left[\begin{array}{cc} R & H \\ \diagdown C=O \quad O=C \diagup \\ H & H \end{array}\right]$ ⟶ $\left[\begin{array}{c} OH \\ O=C \diagdown \\ OH \end{array}\right]$ ⟶ $\boxed{CO_2 + H_2O}$

$\boxed{\begin{array}{c} R \\ \diagdown C=O \\ OH \end{array}}$

Ozonolysis is a two-step procedure that cleaves alkenes. The oxidizing agent ozone is followed by treatment with zinc metal in aqueous acid to yield aldehydes and ketones as shown below.

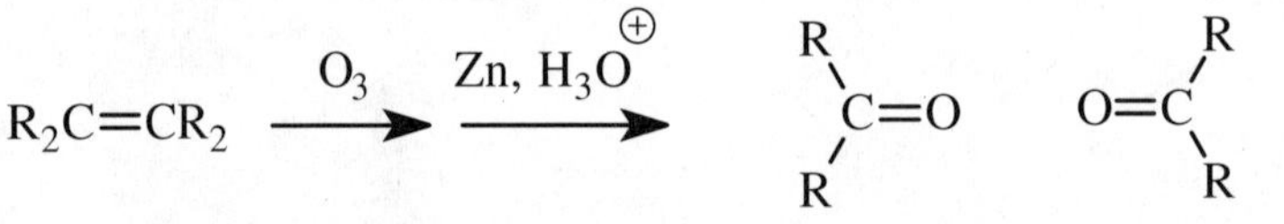

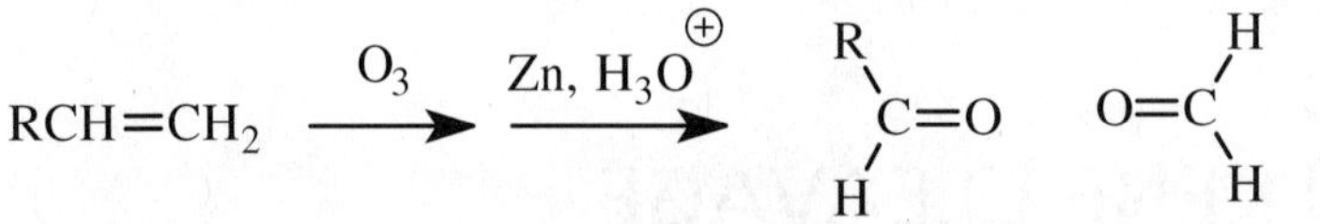

$R_2C=CR_2$ $\xrightarrow{O_3}$ $\xrightarrow{Zn, H_3O^{\oplus}}$ $\begin{array}{cc} R & R \\ \diagdown C=O \quad O=C \diagup \\ R & R \end{array}$

$RCH=CH_2$ $\xrightarrow{O_3}$ $\xrightarrow{Zn, H_3O^{\oplus}}$ $\begin{array}{cc} R & H \\ \diagdown C=O \quad O=C \diagup \\ H & H \end{array}$

When alkenes are treated with osmium (VIII) oxide followed by aqueous sodium bisulfite, vicinal diols result (*syn* addition). These diols can be cleaved to aldehydes and ketones by periodic acid. This three-step sequence is an alternative to ozonolysis.

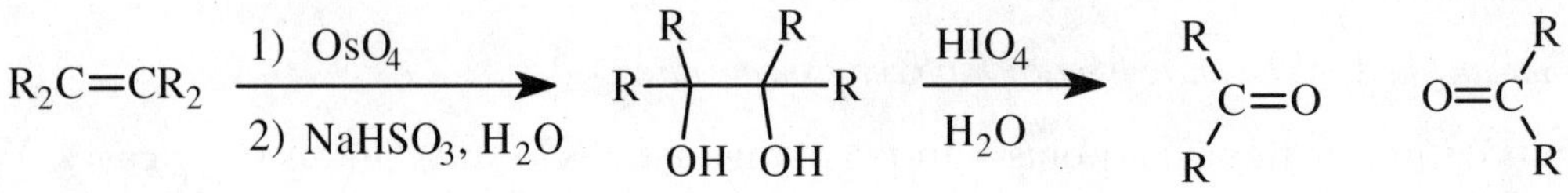

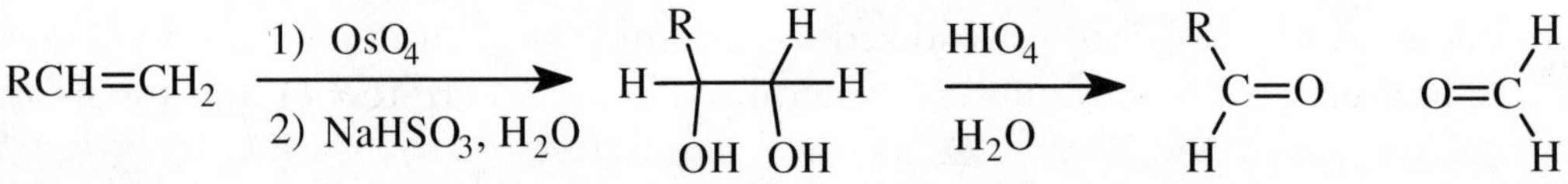

$R_2C=CR_2$ $\xrightarrow[2) NaHSO_3, H_2O]{1) OsO_4}$ $\begin{array}{c} RR \\ R-C-C-R \\ OHOH \end{array}$ $\xrightarrow[H_2O]{HIO_4}$ $\begin{array}{cc} R & R \\ \diagdown C=O \quad O=C \diagup \\ R & R \end{array}$

$RCH=CH_2$ $\xrightarrow[2) NaHSO_3, H_2O]{1) OsO_4}$ $\begin{array}{c} RH \\ H-C-C-H \\ OHOH \end{array}$ $\xrightarrow[H_2O]{HIO_4}$ $\begin{array}{cc} R & H \\ \diagdown C=O \quad O=C \diagup \\ H & H \end{array}$

Topic Test 7: Alkene Cleavage

True/False

1. The final products of alkene cleavage by acidic permanganate are aldehydes and ketones.

2. Treating an alkene with osmium (VIII) oxide followed by aqueous sodium bisulfite will give alkene cleavage products similar to those obtained from ozonolysis.

Multiple Choice

3. Which of the following could not be isolated from an alkene reacting with ozone and then zinc in acid?
   a. Formaldehyde, O=CH$_2$
   b. A ketone
   c. A carboxylic acid
   d. All of the above
   e. None of the above

4. Treatment of a 1,2 diol (vicinal diol) with periodic acid results in
   a. cleavage to alkenes and alkanes.
   b. cleavage to carbon dioxide and water.
   c. cleavage to carboxylic acids and alcohols
   d. All of the above
   e. None of the above

Short Answer

5. Complete the following reaction scheme with unambiguous structural formulas for the missing organic product.

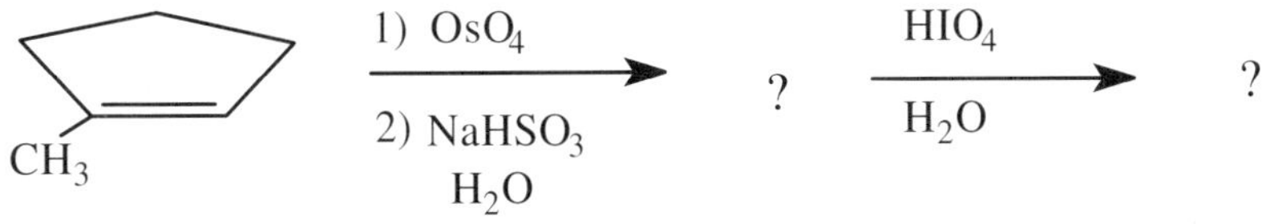

   [1-methylcyclopentene] →(1) OsO$_4$; 2) NaHSO$_3$, H$_2$O)→ ? →(HIO$_4$, H$_2$O)→ ?

6. Complete the following chemical equations with an unambiguous structural formula for the missing organic reactant.

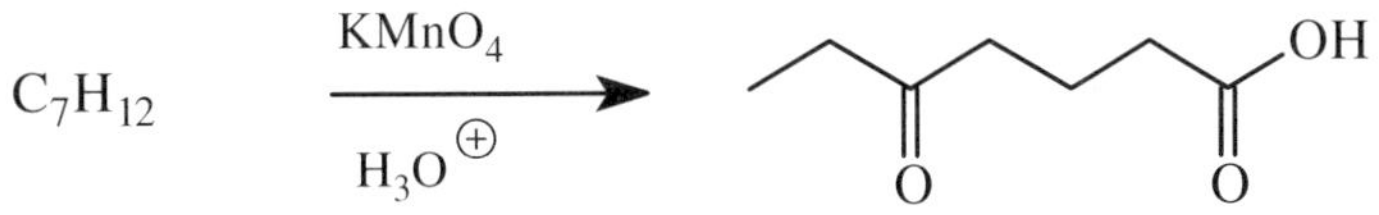

   C$_7$H$_{12}$ →(KMnO$_4$, H$_3$O$^+$)→ [keto-carboxylic acid structure]

Topic Test 7: Answers

1. **False.** Although a disubstituted alkene carbon will yield a ketone, a monosubstituted alkene carbon gives an aldehyde as a fleeting intermediate that is unstable under these reaction conditions and continues to oxidize yielding a carboxylic acid.

2. **False.** The sequence osmium (VIII) oxide followed by aqueous sodium bisulfite will convert an alkene into a vicinal or 1,2 diol, that is, a double alcohol in which the two OH

groups are on what were formerly the alkene carbons. The mode of addition is *syn* so cycloalkenes give *cis* diols.

3. **c.** The ozonolysis conditions listed yield aldehydes and ketones but not carboxylic acids.
4. **e.** Periodic acid cleavage of vicinal diols gives aldehydes and ketones.
5.
6.

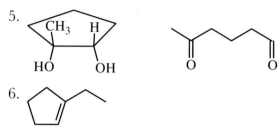

TOPIC 8: RADICAL ADDITIONS OF HBr TO ALKENES

KEY POINTS

✓ *How do peroxides effect the HBr addition to alkenes?*
✓ *What is the mechanism for HBr addition to alkenes when peroxide is present?*
✓ *What products are obtained from alkene treatment with HBr/peroxide?*

When addition of HBr to an alkene is carried out in the presence of trace peroxide, the mechanism is not the ionic reaction shown in Topic 3 but rather proceeds via a **radical chain pathway** illustrated below for propene.

Initiation

1) $RO-OR \longrightarrow 2\ RO^\bullet$

2) $RO^\bullet + CH_2=CHCH_3 \longrightarrow ROCH_2\overset{\bullet}{C}HCH_3$

3) $ROCH_2\overset{\bullet}{C}HCH_3 + HBr \longrightarrow ROCH_2CH_2CH_3 + Br^\bullet$

Propagation

4) $Br^\bullet + CH_2=CHCH_3 \longrightarrow BrCH_2\overset{\bullet}{C}HCH_3$

5) $BrCH_2\overset{\bullet}{C}HCH_3 + HBr \longrightarrow BrCH_2CH_2CH_3 + Br^\bullet$

Then 4) 5) 4) 5) ... etc., until

Termination

6) Several possible termination steps

Because radical stability order is 3° > 2° > 1°, the radical formed in step 4 is secondary and not primary. The net result is non-Markovnikov addition of HBr. Peroxides do not effect the outcome of HCl or HI additions to alkenes.

Topic Test 8: Radical Additions of HBr to Alkenes

True/False

1. The presence of peroxides alters the mechanism by which HBr adds to alkenes.
2. When HCl or HI are added to alkenes in the presence of peroxides, the resulting products are predictable according to Markovnikov's rule.

Multiple Choice

3. In the radical addition of HBr addition to alkenes peroxide serves as a
   a. catalyst
   b. intermediate
   c. initiator
   d. All of the above
   e. None of the above

4. Which alkene below would give different products for HBr addition depending on whether or not peroxide was present?
   a. cyclohexene
   b. *trans*-3-hexene
   c. 2,3-dimethyl-2-butene
   d. All of the above
   e. None of the above

Short Answer

5. Complete the following chemical equations with an unambiguous structural formula for the missing organic product.

$$\text{methylenecyclopentane} \xrightarrow[\text{Peroxide}]{\text{HBr}}$$

6. Complete the following chemical equation with an unambiguous structural formula for the missing organic reactant.

$$\underline{\qquad} \xrightarrow[\text{Peroxides}]{\text{HBr}} \text{CH}_3\text{CH}(\text{CH}_3)\text{CH}(\text{Br})\text{CH}_2\text{CH}_3$$

Topic Test 8: Answers

1. **True.** When peroxides are present, the addition proceeds via a radical chain mechanism, whereas without peroxides, the mechanism is ionic via a carbocation intermediate.

2. **True.** Peroxides do not change the outcome of an addition of HCl or HI to an alkene, and Markovnikov's rule allows us to predict that the product will have the halogen on the more substituted of what were formerly alkene carbons.

3. **c.** A trace of peroxide supplies the first free radicals that start the chain reaction.

4. **e.** The presence of peroxides causes the product to be non-Markovnikov versus the Markovnikov products observed without peroxide. Because all alkenes named in a, b, and c are symmetrical, it would not matter which way the HBr adds across the alkene and the same product results either way.

5. 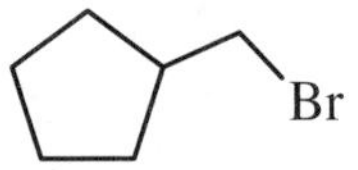 [cyclopentylmethyl bromide structure]

6. 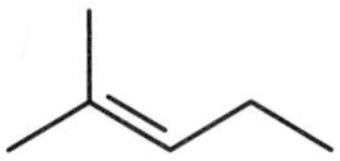 [2-methyl-2-butene structure]

TOPIC 9: ALKENE POLYMERIZATION

Key Points

✓ *What are alkene polymers?*

✓ *What polymer will result from the polymerization of a given alkene?*

✓ *How are alkene polymers named?*

✓ *How are the structures of polymers represented?*

Under some circumstances, alkenes react with themselves to form large molecules called **polymers**. The reactions can be cationic or radical, but the latter are more important in industry. Radical polymerization normally requires some small amount of initiator such as peroxide to start the chain reaction. The chains vary in size but are usually long. The names often have the form "poly____," where the blank is the **monomer** name. Several important alkene polymers are represented below. Structural formulas for polymers are usually represented by enclosing the repeating unit within parentheses and using a subscript "n" as shown below.

Monomer		Polymer	
$CH_2=CH_2$	Ethene (ethylene)	$-(CH_2CH_2)_n-$	Polyethylene
$CH_2=CHCH_3$	Propene (propylene)	$-(CH_2CH)_n-$ $\quad\quad\quad\;\, |$ $\quad\quad\quad CH_3$	Polypropylene
$CH_2=CHCl$	Vinyl chloride	$-(CH_2CH)_n-$ $\quad\quad\quad\;\, |$ $\quad\quad\quad Cl$	Polyvinyl chloride
$CH_2=CH$ $\quad\quad\;\, |$ $\quad\quad Ph$	Styrene	$-(CH_2CH)_n-$ $\quad\quad\quad\;\, |$ $\quad\quad\quad Ph$	Polystyrene

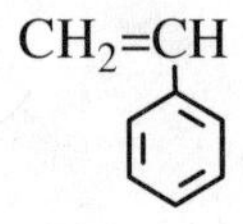

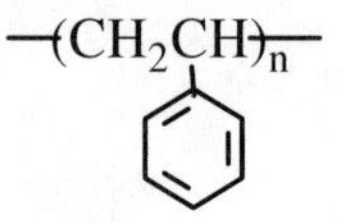

Topic Test 9: Alkene Polymerization

True/False

1. Alkene polymerization reactions can occur by carbocation or radical mechanisms.

2. The repeating unit (the part within parentheses) in an alkene polymer structural formula should have the same molecular formula as the monomer from which the polymer was synthesized.

Multiple Choice

3. A polymer with structure $(CF_2CF_2)_n$ could be named
   a. vinyl fluoride
   b. polyvinyl fluoride
   c. tetrafluoroethene
   d. All of the above
   e. None of the above

4. The "n" subscript outside the parentheses in a structural formula representation of a polymer denotes
   a. one or two
   b. no more than three
   c. at least 10,000
   d. a large but imprecise number of repeating units
   e. None of the above

Short Answer

5. Complete the following reaction scheme with an unambiguous structural formula for the missing organic product.

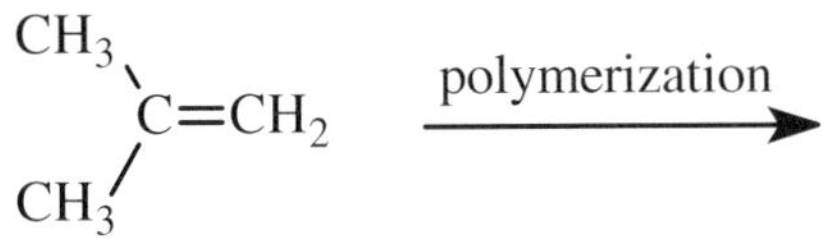

$$\underset{\text{Isobutylene}}{\underset{CH_3}{\overset{CH_3}{>}}C=CH_2} \xrightarrow{\text{polymerization}} \underset{\text{Polyisobutylene}}{}$$

6. Complete the following reaction scheme with an unambiguous structural formula for the missing organic reactant.

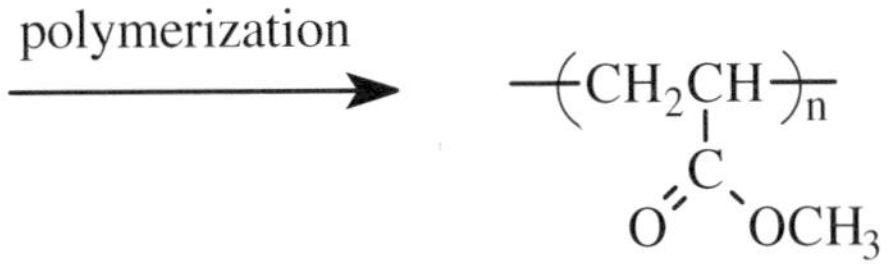

$$\xrightarrow{\text{polymerization}} \underset{\text{Poly(methyl acrylate)}}{-(CH_2CH)_n \text{ with } -C(=O)OCH_3 \text{ substituent}}$$

Topic Test 9: Answers

1. **True.** Both mechanisms are possible, but radical polymerization of alkenes is more important in industry.

2. **True.** There are no atoms lost or gained in the propagation steps of an alkene polymerization reaction and all the atoms of the monomer will appear in the repeating unit.

3. **e.** The structure seems to be a polymer of tetrafluoroethene so the name would reasonably be polytetrafluoroethene. Parentheses are sometimes used in the name, and the parent term "ethylene" is used in place of "ethene" by many industrial chemists. Thus, names like "poly(tetrafluoroethylene)" are acceptable. Response a is the name for $CH_2=CHF$. Response b is the polymer $-(CH_2CHF)_n-$, and response c is the name of the monomer $CF_2=CF_2$.

4. **d.** A polymer, by definition, must be made of many individual parts ("poly" = many + "mer" = parts), and several different polymer chains will likely have different numbers of repeating units.

5. 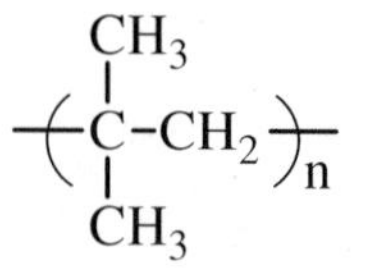 $-(C(CH_3)_2-CH_2)_n-$

6. 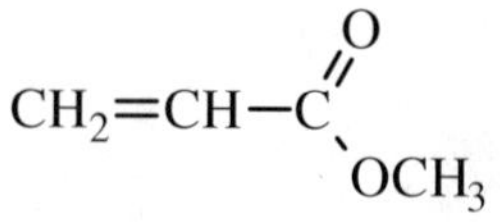 $CH_2=CH-C(=O)OCH_3$ Methyl acrylate

TOPIC 10: PREPARATION OF ALKENES: ELIMINATION REACTIONS

KEY POINTS

✓ *What is an elimination reaction?*

✓ *What are two common categories of alkene-forming elimination reactions?*

✓ *What reagents and/or conditions are used for dehydration?*

✓ *What reagents and/or conditions are used for dehydrohalogenation?*

✓ *What is Zaitsev's rule?*

Alkenes are often prepared by reactions that formally resemble the reverse of electrophilic additions. These are called elimination reactions, and they involve the loss of atoms or groups from neighboring carbons and result in the formation of pi bonds.

Two important elimination reactions are dehydration (H_2O loss) and dehydrohalogenation (HX loss). The first of these is carried out on alcohols usually in the presence of concentrated sulfuric or phosphoric acid. Unless carbocation rearrangements take place, the products are predictable and straightforward. Alternatively, the reaction is sometimes performed with phosphorous oxychloride ($POCl_3$). For cases where more than one alkene can be formed, the major product is usually the one that has the most substitution around the alkene linkage (i.e., $R_2C=CR_2$ > $RHC=CR_2$ > $RHC=CHR \cong H_2C=CR_2$ > $H_2C=CHR$). This observation is known as **Zaitsev's rule** and it also holds for HX elimination. The most common dehydrohalogenation reagents are bases (KOH, NaOR, NR_3, etc.). The mechanisms for these reactions are discussed in Chapter 6. Some representative elimination reactions are shown below.

Topic Test 10: Preparation of Alkenes: Elimination Reactions

True/False

1. Elimination reactions convert alkenes into alcohols and alkyl halides.

2. Zaitsev's rule predicts that an elimination reaction will yield the least substituted alkene.

Multiple Choice

3. Which reagents/conditions are *not* used for dehydration reactions?
   a. Concentrated sulfuric acid
   b. Concentrated phosphoric acid
   c. Phosphorus oxychloride
   d. KOH/ethanol
   e. All of the above

4. Which reaction below is an elimination?
   a. Hydroboration
   b. Oxymercuration

c. Hydrogenation
d. Dehydrohalogenation
e. All of the above

Short Answer

5. Complete the following scheme with unambiguous structural formulas and all possible dehydration products. Indicate which product will be the major isomer formed.

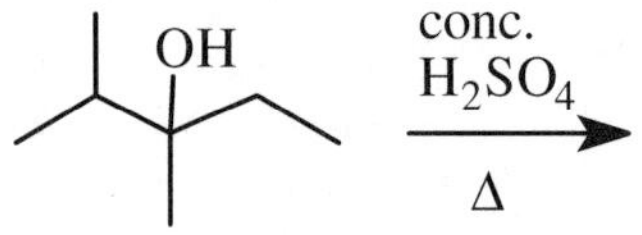

6. Complete the following scheme with an unambiguous structural formula for the missing organic reactant.

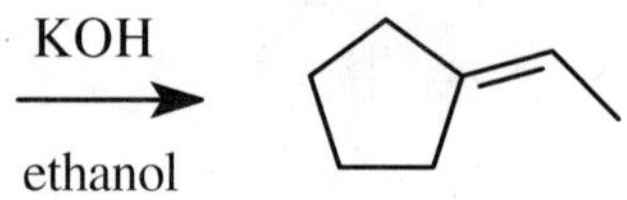

Topic Test 10: Answers

1. **False.** Actually the opposite is true, that is, alkyl halides and alcohols undergo elimination reactions to yield alkenes.

2. **False.** Zaitsev's rule predicts the elimination reaction gives the *most* substituted alkene.

3. **d**

4. **d.** As the name suggests, this reaction eliminates HX.

5.

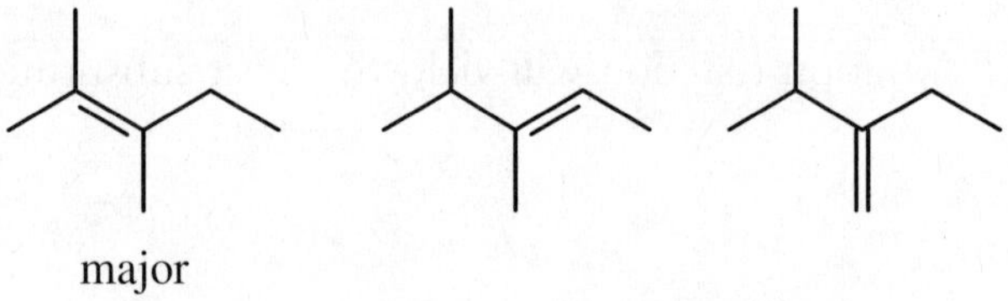

6.

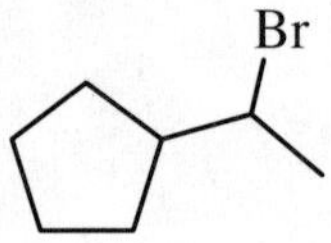

APPLICATION

Many of the common industrial materials we see every day are alkene polymers. They are often better known by their trade names. Some of these are shown below.

Trade Name	Chemical Name	Structure	Uses
Styrofoam	Polystyrene	$-(CH_2CH)_n-$ with phenyl substituent	Insulation, packing
Teflon	Polytetrafluoroethene	$-(CF_2CF_2)_n-$	Non-stick surfaces
Plexiglass or Lucite	Poly(methyl methacrylate)	$-(CH_2C(CH_3)(CO_2CH_3))_n-$	Glass substitute
Saran	Poly(1,1-dichloroethene)	$-(CH_2CCl_2)_n-$	Food packaging

DEMONSTRATION PROBLEM

An unidentified compound was treated by ozone followed by zinc in acid. The resulting product mixture contained equimolar amounts of a three-carbon acyclic ketone, a three-carbon aldehyde, and a two-carbon dialdehyde. Propose a structure for the unidentified compound.

Solution

First, draw structures for the products from the descriptions given. There is only one possibility for each of the three products described.

$$CH_3CCH_3 \quad H-C(=O)-C(=O)-H \quad CH_3CH_2CH(=O)$$

$$\text{3-carbon ketone} \qquad \text{2-carbon dialdehyde} \qquad \text{3-carbon aldehyde}$$

Now recall that the ozonolysis reaction sequence converts what were alkene carbons into aldehyde and ketone carbons. Every place there is a carbonyl in the products represents where an alkene linkage was in the original molecule. There is only one product that has two carbonyl groups so it must have been in the center. The other two products came from alkene fragments that were attached to either side as shown.

$$(CH_3)_2C=CH-CH=CH-CH_2CH_3 \xrightarrow{\text{1) } O_3}{\text{2) Zn, } H_3O^{\oplus}} (CH_3)_2C=O + O=CH-CH=O + O=CH-CH_2CH_3$$

Both E and Z are reasonable structures.

Topic 10: Preparation of Alkenes: Elimination Reactions

Chapter Test

True/False

1. There are eight hydrogens in a molecule of 1,3-cyclohexadiene.

2. Treatment of cyclohexene with aqueous mercury (II) acetate followed by sodium borohydride produces the same alcohol product obtained by treating cyclohexene with borane then alkaline hydrogen peroxide.

3. Oxygen can be ignored when calculating the degree of unsaturation.

4. Nucleophiles generally have a positive or partial positive charge.

5. Addition of HBr to alkenes in the presence of trace peroxide yields non-Markovnikov products via radical chain mechanism.

Multiple Choice

6. Compound **A** has the formula $C_{10}H_{16}$. What is the degree of unsaturation for **A**?
   a. 1
   b. 2
   c. 3
   d. 4
   e. None of the above

7. Compound **A** has the formula $C_{10}H_{16}$ and reacts with excess hydrogen over palladium to yield compound **B** with the formula $C_{10}H_{18}$. How many rings are in **A**?
   a. 1
   b. 2
   c. 3
   d. 4
   e. None of the above

8. An acyclic molecule that has no pi bonds and contains "n" carbons and one nitrogen will have how many hydrogen atoms?
   a. n + 1
   b. 2n
   c. 2n + 2
   d. 2n + 3
   e. None of the above

9. The notation used for representing a reaction mechanism requires that curved arrows
   a. indicate the direction of atomic motion.
   b. show the direction of electron flow (i.e., nucleophile attacking electrophile).
   c. show the direction of positive charge flow (i.e., electrophile attacking nucleophile).
   d. All of the above
   e. None of the above

10. Which steps below will convert an alkene into an alcohol?
    a. Water and acid catalyst
    b. Water and mercury acetate followed by sodium borohydride
    c. Borane followed by alkaline hydrogen peroxide

d. All of the above
e. None of the above

11. During the reaction of an alkene with Br_2/H_2O to yield a halohydrin, the nucleophile is
    a. Br—Br
    b. bromide ion
    c. hydroxide ion
    d. water
    e. None of these

Short Answer and Essay

12. When vinylcyclobutane is treated with HBr/ether, the product mixture contains the three compounds shown below. Offer a mechanistic explanation for this observation.

13. Provide an unambiguous structural formula for Z-3-methyl-1,3,5-hexatriene.

14. Draw and name the monomers for each of the materials listed in APPLICATION (p. 67).

15. Name the following compound.

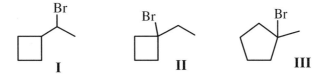

16. + HBr → ?

17. ? $\xrightarrow{BH_3}$ $\xrightarrow{H_2O_2, \ominus OH}$ —OH

18. $\xrightarrow{Br_2, H_2O}$?

19. $\xrightarrow[PtO_2]{excess\ H_2}$?

20. ? $\xrightarrow{O_3}$ $\xrightarrow[Zn]{H_3O^\oplus}$ + $CH_3CH_2CH_2\overset{O}{\underset{\|}{C}}H$

Chapter 3 Test: Answers

1. **T** 2. **T** 3. **T** 4. **F** 5. **T** 6. **c** 7. **b** 8. **d** 9. **b** 10. **d** 11. **d**

12. The starting material is protonated to give a secondary carbocation that has three possible fates. First, it could be captured by the bromide nucleophile to yield the simple Markovnikov product **I**. Second, it could rearrange by a 1,2-hydride shift to a tertiary carbocation that is captured by bromide to yield **II**. Third, it could rearrange by a 1,2 alkyl shift to another secondary carbocation that has less ring strain. (The "alkyl shift" involves a ring atom and therefore appears as a ring expansion.) That carbocation is trapped by the bromide to give **III**.

13.

14. $CH_2=CH$-(phenyl) $CF_2=CF_2$ $CH_2=C(CH_3)-CO_2CH_3$ $CH_2=CCl_2$

    Styrene   Tetrafluoroethene   Methyl methacrylate   1,1-dichloroethene

15. *E*-6-Methyl-1,3-octadiene

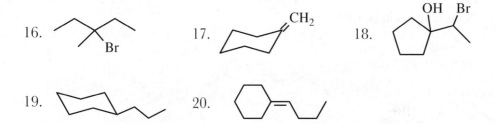

19.   20.

Check Your Performance

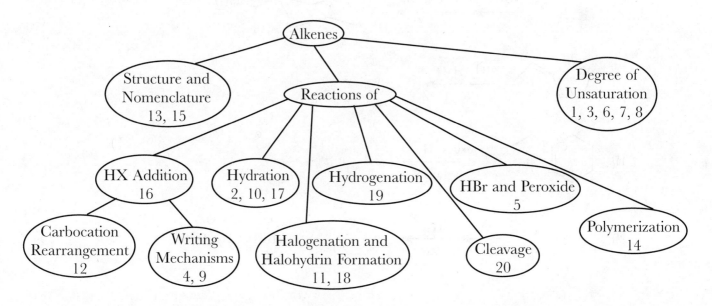

CHAPTER 4

Alkynes

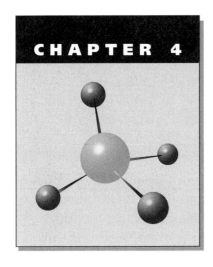

The flame of a welder's oxyacetylene torch results from the combustion (Chapter 2) of acetylene (ethyne, H—C≡C—H), the simplest alkyne. In this chapter we survey the chemistry of alkynes. There are relatively few naturally occurring alkynes, and most of those contain other functional groups. The same is true of the synthetic pharmaceuticals that contain carbon-carbon triple bonds.

ESSENTIAL BACKGROUND

- Orbitals and hybridization (Chapter 1)
- Acid and bases (Chapter 1)
- Drawing structural formulas (Chapter 1)
- Functional groups: alkene, alkyne, alcohol, aldehyde, ketone, carboxylic acid (Chapter 2)
- Alkene nomenclature (Chapter 3)
- Electrophilic addition to pi bonds (Chapter 3)
- Markovnikov's rule (Chapter 3)

TOPIC 1: STRUCTURE, NOMENCLATURE, AND PREPARATION

KEY POINTS

✓ *What is the hybridization and shape around alkyne carbons?*

✓ *How are alkynes named?*

✓ *How are alkynes generally prepared?*

Alkynes contain a carbon-carbon triple bond. The triple-bonded carbons are sp hybridized, and the shape is linear. The simplest alkyne, ethyne, is shown below.

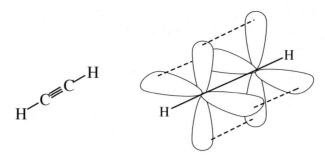

Ethyne is perhaps better known by its common name, acetylene. Alkynes are sometimes collectively referred to as **acetylenes**. The International Union of Pure and Applied Chemistry (IUPAC) names are similar to those for alkenes in that the longest chain containing the triple bond is considered the parent and a number is used to specify the first carbon where the triple bond is encountered. The parent chain is numbered from the end that gives the triple-bonded carbons the lowest numbers possible. No *cis/trans* is possible around the linear alkyne linkage. As usual, alkyl substituents are mentioned as prefixes with numbers to specify their location. Multiple alkyne linkages are specified as diynes, triynes, and so on. The following examples illustrate these conventions:

Propyne	$CH_3C \equiv CH$
1-Butyne	$CH_3CH_2C \equiv CH$
2-Butyne	$CH_3C \equiv CCH_3$
2-Pentyne	$CH_3CH_2C \equiv CCH_3$
1,3-Pentadiyne	$HC \equiv C-C \equiv CCH_3$
4-Methyl-2-pentyne	$(CH_3)_2CHC \equiv CCH_3$
2,2,7-Trimethyl-4-octyne	$(CH_3)_3CCH_2C \equiv CCH_2CH(CH_3)_2$

Alkynes are generally prepared via elimination reactions similar to those we saw in Chapter 3 (Topic 10). For example, a double dehydrohalogenation of a dihalo compound will often yield an alkyne as shown below.

$$R-\underset{\underset{H}{|}}{\overset{\overset{X}{|}}{C}}-\underset{\underset{H}{|}}{\overset{\overset{X}{|}}{C}}-R \xrightarrow[\text{(-HX)}]{\text{excess NaNH}_2} \left[R-\underset{\underset{H}{|}}{\overset{\overset{X}{|}}{C}}=C-R \right] \xrightarrow{\text{(-HX)}} R-C \equiv C-R$$

(X = Cl, Br, I)

Topic Test 1: Structure, Nomenclature, and Preparation

True/False

1. The "ene" suffix on the name acetylene indicates it is an alkene.
2. All four carbons of 2-butyne are on the same axis.

Multiple Choice

3. Which statement is true about alkynes?
   a. They have at least two degrees of unsaturation.
   b. They can be prepared via elimination reactions.
   c. Their systematic IUPAC names end in "yne."

d. All of the above
e. None of the above

4. The formula for 3,6-diethyl-1-octyne is
   a. C_8H_{16}
   b. $C_{10}H_{22}$
   c. $C_{12}H_{24}$
   d. $C_{12}H_{20}$
   e. None of the above

Short Answer

5. Name the following compound

$$CH_3CH_2C\equiv C-\underset{\underset{\text{cyclopentyl}}{|}}{\overset{\overset{CH_3}{|}}{C}}-CH_3$$

6. Draw an unambiguous structural formula for 3,3-dimethyl-1,4-hexadiyne.

Topic Test 1: Answers

1. **False.** Although it is often true that systematic names for alkenes end with an "ene," the name "acetylene" is actually a common name for ethyne, which is the two-carbon alkyne.

2. **True.** The triple-bonded inside carbons of 2-butyne are each sp hybridized and have a linear shape (bond angle = 180 degrees). This requires that they, as well as whatever atoms are attached directly to them, are colinear.

3. **d**

4. **e.** A diethyl octyne is an eight-carbon chain with two appendages of two carbons each. That gives a total of 12 carbons. The name indicates it is an acyclic alkyne so the degree of unsaturation is expected to be two. For a 12-carbon compound that would mean 22 hydrogens, and the formula would be $C_{12}H_{22}$.

5. 2-Cyclopentyl-2-methyl-3-hexyne.

6. $$HC\equiv C-\underset{\underset{CH_3}{|}}{\overset{\overset{CH_3}{|}}{C}}-C\equiv C-CH_3$$

TOPIC 2: SOME ELECTROPHILIC ADDITIONS

KEY POINTS

✓ *What products result from addition of halogens to alkynes?*

✓ *What products result from addition of HX to alkynes?*

Many reactions of alkynes are similar to those for alkenes. Two examples are halogenation and HX addition. In the presence of excess chlorine or bromine, alkyne will add two equivalents of halogen as shown below. The reaction occurs via a *trans* dihaloalkene that is not normally isolated but instead reacts quickly with the second equivalent of molecular halogen to yield a tetrahalogenated compound.

$$R-C\equiv C-R' \xrightarrow{X_2} \left[\begin{array}{c} X \\ \diagdown \diagup R' \\ C=C \\ \diagup \diagdown \\ R X \end{array} \right] \xrightarrow{X_2} R-\underset{\underset{X}{|}}{\overset{\overset{X}{|}}{C}}-\underset{\underset{X}{|}}{\overset{\overset{X}{|}}{C}}-R'$$

R and R' = H or alkyl
X = Cl or Br

Two equivalents of HX will add to an alkyne. The regiochemistry is usually predictable using Markovnikov's rule if the alkyne is terminal. Unsymmetrical internal alkynes usually yield a mixture of products as shown below.

$$R-C\equiv C-H \xrightarrow{HX} R-\overset{\overset{X}{|}}{C}=CH_2 \xrightarrow{HX} R-\underset{\underset{X}{|}}{\overset{\overset{X}{|}}{C}}-CH_3$$

$$R-C\equiv C-R' \xrightarrow{HX} R-\underset{\underset{X}{|}}{C}=CHR' + R-CH=\underset{\underset{X}{|}}{C}-R'$$

Topic Test 2: Some Electrophilic Additions

True/False

1. Alkynes react with HX and X-X to yield addition products.
2. Terminal alkynes yield 1,1-dibromo compounds when reacted with two equivalents of HBr.

Multiple Choice

3. Which of the following would likely yield a mixture of products when reacted with one equivalent of HBr? (ignore *cis/tans*)
   a. 1-hexyne
   b. 2-hexyne
   c. 3-hexyne
   d. All of the above
   e. None of the above

4. Which product below could be made by the addition of bromine to an alkyne?
   a. $CH_3CH_2CHBr_2$
   b. $CH_3CBr_2CH_3$
   c. $CH_3CBr_2CHBr_2$
   d. All of the above
   e. None of the above

Short Answer

5. Provide unambiguous structural formulas for the missing organic products.

cyclopentyl-C≡CCH₂CH₃ →(1 equiv. HBr)

6. Provide an unambiguous structural formula for the missing organic reactant.

→(excess HBr) CH₃CBr₂-(cyclohexane)-CBr₂CH₃

Topic Test 2: Answers

1. **True.** All the examples of this section are addition reactions of this type.
2. **False.** The dibromo product that results from two equivalents of HBr adding to a terminal alkyne would have Markovnikov orientation and therefore would be 2,2-dibromo.
3. **b.** Because this is an internal unsymmetrical alkyne, there are two possible orientations that HBr can have with respect to the alkyne. The product mixture would likely contain 2-bromo-2-hexene and 3-bromo-2-hexene. Addition of HBr to 3-hexyne can take place in either direction, but both paths lead to the same product.
4. **c.** This product came from adding two equivalents of bromine to propyne. Choices a and b contain only two bromine atoms each and could not have resulted from halogenation of an alkyne.
5. cyclopentyl-C(Br)=CHCH₂CH₃ (E or Z) + cyclopentyl-CH=C(Br)CH₂CH₃ (E or Z)
6. H-C≡C-(cyclohexane)-C≡C-H

TOPIC 3: HYDRATION OF ALKYNES: KETO-ENOL TAUTOMERISM

KEY POINTS

✓ What reagents are used for Markovnikov hydration of alkynes?

✓ What is the structure of an enol?

✓ What keto tautomer is in equilibrium with a given enol?

✓ Which tautomer is usually favored in keto-enol tautomerism?

✓ What products will be isolated from the hydration reaction of an alkyne?

✓ What products result from hydroboration of alkynes?

Like alkenes, alkynes can add water across their pi bonds. The addition of one equivalent of water produces a molecule that is simultaneously an alkene and an alcohol. Compounds of this type where the alcohol OH is attached to one of the alkene carbons are called **enols**. Under most circumstances these enol intermediates are not stable and they spontaneously isomerize to structures in which the pi bond and the alcohol hydrogen are formally moved to yield a carbonyl compound. This is known as **tautomerism** and the isomers are called **tautomers**. The carbonyl tautomer is often (though not always) a ketone and is called the **keto** tautomer. It is normally favored over the enol tautomer in an equilibrium. The net transformation resulting from alkyne hydration is conversion of an alkyne into a carbonyl compound. The reagent commonly used here is water in the presence of mercuric sulfate and sulfuric acid. The hydration of alkynes is useful for making ketones provided the starting alkyne is symmetrical or terminal; otherwise, mixtures result as summarized below.

$$R-C\equiv C-H \xrightarrow[H_2SO_4]{H_2O, HgSO_4} \left[R-\underset{OH}{C}=CH_2 \right] \rightleftarrows R-\underset{O}{\overset{\|}{C}}-CH_3$$

Terminal alkyne — Markovnikov enol — A methyl ketone

$$R-C\equiv C-R \xrightarrow[H_2SO_4]{H_2O, HgSO_4} \left[R-\underset{OH}{C}=CHR \right] \rightleftarrows R-\underset{O}{\overset{\|}{C}}-CH_2R$$

Symmetrical alkyne — A ketone

$$R-C\equiv C-R' \xrightarrow[H_2SO_4]{H_2O, HgSO_4} \left[\begin{array}{c} R-\underset{OH}{C}=CHR' \\ + \\ R-CH=\underset{OH}{C}R' \end{array} \right] \rightleftarrows \begin{array}{c} R-\underset{O}{\overset{\|}{C}}-CH_2R' \\ + \\ R-CH_2-\underset{O}{\overset{\|}{C}}R' \end{array}$$

Unsymmetrical alkyne — Two different ketones

As with alkynes, hydroboration of alkynes yields the non-Markovnikov products. Terminal alkynes will lead to aldehydes under these conditions. Internal alkynes yield mixtures of ketones unless the alkyne is symmetrical. The alkyne reacts with two B—H bonds to yield an intermediate that, upon treatment with alkaline hydrogen peroxide, produces a hydrate that in turn loses water to give the carbonyl compound. This is illustrated below for conversion of a terminal alkyne to an aldehyde.

$$R-C\equiv C-H \xrightarrow{BH_3} \left[\begin{array}{c} H \quad B \\ | \quad | \\ R-C-CH \\ | \quad | \\ H \quad B \end{array} \right] \xrightarrow[H_2O_2]{OH^\ominus} \left[R-CH_2-CH\underset{OH}{\overset{OH}{\diagup}} \right] \rightleftarrows \begin{array}{c} O \\ \| \\ RCH_2CH \\ + H_2O \end{array}$$

76 Chapter 4 Alkynes

Topic 3 Test: Hydration of Alkynes: Keto-Enol Tautomerism

True/False

1. The addition product normally isolated from hydration of an alkyne is an enol.

2. The hydration of a terminal alkyne with water and mercuric sulfate will yield an aldehyde as the final product.

Multiple Choice

3. Which of the following would yield a mixture of ketone products upon hydration?
   a. 1-Hexyne
   b. 2-Hexyne
   c. 3-Hexyne
   d. All of the above
   e. None of the above

4. Treating a terminal alkyne with borane followed by alkaline hydrogen peroxide will yield
   a. an enol.
   b. a ketone.
   c. an aldehyde.
   d. an alcohol.
   e. None of the above

Short Answer

5. Draw the "keto" tautomer of the "enol" shown.

$$CH_3-\underset{\underset{CH_3}{|}}{\overset{\overset{CH_3}{|}}{C}}-\underset{}{\overset{\overset{OH}{|}}{C}}=CHCH_3 \quad \rightleftharpoons$$

6. What reagents and/or conditions could one use to carry out the following conversion? More than one step may be required.

Topic Test 3: Answers

1. **False.** Although the intermediate formed in most alkyne hydrations is an enol, it is not isolated from the reaction mixture because enols normally rearrange to the corresponding keto tautomers.

2. **False.** When the hydration of a terminal alkyne is carried out using mercuric sulfate, the enol intermediate will have Markovnikov orientation and therefore the OH will be on the inside carbon (C2). That enol intermediate will tautomerize to the favored keto tautomer, which will be a methyl ketone (a ketone with a methyl on the carbonyl).

3. **b.** This unsymmetrical internal alkyne will yield a mixture of ketones in which the carbonyl is the second or third carbon of a six-carbon chain. Because a is a terminal alkyne, we expect only one (Markovnikov) product, and because c is a symmetrical alkyne, it does not matter which orientation the addition of water takes.

4. **c.** These reagents produce a 1,1-dialcohol RCH$_2$CH(OH)$_2$, called a hydrate, that spontaneously dehydrates to give an aldehyde.

5.
$$(CH_3)C\overset{\overset{\displaystyle O}{\|}}{C}CH_2CH_3$$

6. H$_2$O, H$_2$SO$_4$, HgSO$_4$ *or* 1) BH$_3$ then 2) H$_2$O$_2$, OH$^-$

TOPIC 4: REDUCTION OF ALKYNES TO ALKENES AND ALKANES

KEY POINTS

✓ *Under what conditions can an alkyne be converted to an alkane?*

✓ *Under what conditions can an alkyne be converted to a cis-alkene?*

✓ *Under what conditions can an alkyne be converted to a trans-alkene?*

In the presence of excess hydrogen and a suitable catalyst, alkynes take up two equivalents of hydrogen to yield alkanes. The reaction is difficult to stop at the alkene stage, and other more reliable strategies exist for partial reduction.

$$R-C\equiv C-R' \xrightarrow[\text{Pd/C}]{2\,H_2} \left[\begin{array}{c} R \\ \diagdown \\ C=C \\ \diagup \\ H \end{array} \begin{array}{c} R' \\ \diagup \\ \\ \diagdown \\ H \end{array} \right] \longrightarrow RCH_2CH_2R'$$

Stopping the reduction at the alkene phase is often done with the help of **Lindlar's catalyst**, which is slightly deactivated palladium metal on CaCO$_3$. This catalyst is less effective than Pd/C, and it allows one equivalent of hydrogen to react, yielding a *cis* alkene.

$$R-C\equiv C-R' \xrightarrow[\text{catalyst}]{H_2 \atop \text{Lindlar's}} \begin{array}{c} R \\ \diagdown \\ C=C \\ \diagup \\ H \end{array} \begin{array}{c} R' \\ \diagup \\ \\ \diagdown \\ H \end{array}$$

trans alkenes can be obtained from alkynes by treating them with lithium in liquid ammonia.

$$R-C\equiv C-R' \xrightarrow[\text{NH}_3]{\text{Li}} \begin{array}{c} H \\ \diagdown \\ C=C \\ \diagup \\ R \end{array} \begin{array}{c} R' \\ \diagup \\ \\ \diagdown \\ H \end{array}$$

Topic 4 Test: Reduction of Alkynes to Alkenes and Alkanes

True/False

1. Lindlar's catalyst is used to promote the hydrogenation of alkyne to an alkane.

2. The product of an alkyne reaction with lithium metal in liquid ammonia is a *trans* alkene.

Multiple Choice

3. The reaction of an alkyne with hydrogen gas over palladium metal
   a. produces alkanes.
   b. is a reduction.
   c. is called hydrogenation.
   d. All of the above
   e. None of the above

4. Which reagents and conditions will convert 2-pentyne into *cis*-2-pentene?
   a. H_2 over Pd/C
   b. Li, $NH_3(l)$
   c. H_2O, $HgSO_4$, H_2SO_4
   d. All of the above
   e. None of the above

Short Answer

5. Write out a reaction showing the conversion of 2,5-dimethyl-3-hexyne into *trans*-2,5-dimethyl-3-hexene.

6. Provide an unambiguous structural formula for the product of the following reaction.

$$CH_3CH_2\diagdown_{C=C}\diagup^{CH_2CH_2C\equiv CCH_2CH_3} \xrightarrow[\text{catalyst}]{H_2 \text{ Lindlar's}}$$

Topic Test 4: Answers

1. **False.** Lindlar's catalyst is normally used to partially hydrogenate alkynes to the corresponding *cis* alkenes. Palladium on carbon is the usual catalyst used for hydrogenation to an alkane.
2. **True**
3. **d**
4. **e**
5. $(CH_3)_2CHC\equiv CCH(CH_3)_2 \xrightarrow[NH_3]{Li}$

6.  (*cis*, *cis* diene)

TOPIC 5: ACIDITY OF ALKYNES AND ALKYLATION OF ACETYLIDES

KEY POINTS

✓ *What is the relative acidity of hydrogen on sp^3, sp^2, and sp-hybridized carbon?*

✓ *What reagent is used to remove a proton from a terminal alkyne?*

✓ *What is an acetylide ion?*

✓ *How can an acetylide be alkylated?*

The acidity of a hydrogen on carbon increases as the amount of s character in the bond increases. Because s orbitals are closer to the positively charged nucleus, one might expect electrons in these orbitals to be more tightly held than those of a p orbital. In other words, an orbital with more s character is more electronegative. **Table 4.1** shows several different C—H environments along with the hybridization and the percentage of s and p character in the C—H bonds and approximate pK_a.

Table 4.1 Acidity and Hybridization of C—H Bonds				
Structure	**Hybridization**	**% s**	**% p**	**pK (approx.)**
CH_3CH_3	sp^3	25	75	50
$CH_2=CH_2$	sp^2	33	67	44
H—C≡C—H	sp	50	50	25

Note that although terminal alkynes are quite weak acids by most standards, they are clearly more acidic than other types of organic molecules we have discussed. They can be deprotonated if a strong enough base is used. Hydroxide and some other common bases are too weak, but the amide ion of $Na^+NH_2^-$ is sufficient as shown below. Ammonia (pK_a about 38) is less acidic than acetylene. Therefore, one predicts that the NH_2^- ion is a stronger base than H—C≡C:⁻

$$R-C\equiv C-H \ + \ ^-NH_2 \ \longrightarrow \ R-C\equiv C:^- \ + \ NH_3$$

The conjugate base of a terminal alkyne is called an **acetylide**. These anionic species are useful for building carbon-carbon bonds. When reacted with a methyl or primary alkyl halides, acetylides are alkylated as shown below. The mechanism for this process is described more fully in Chapter 6, but for now notice that the acetylide is a nucleophile that attacks the back side of the carbon that bears the halogen. If the alkyl halide is not methyl or primary, the back side is too crowded to allow this approach, and the reaction will instead yield alkenes via an elimination mechanism.

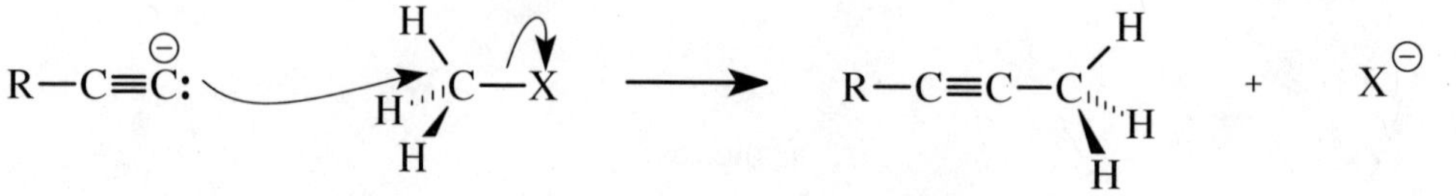

Methyl halide → Alkylation (substitution)

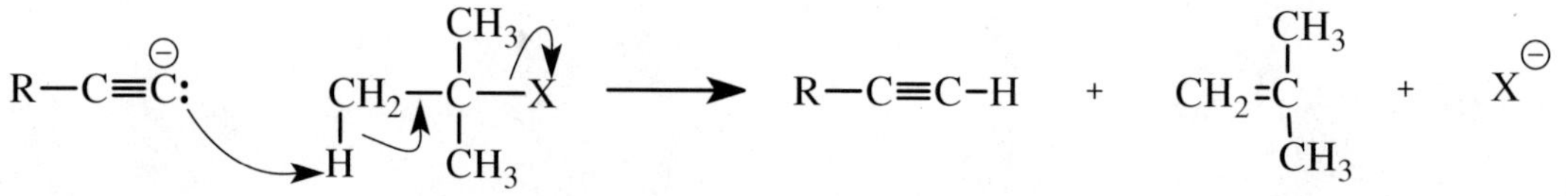

Tertiary halide → Alkene (elimination)

80   Chapter 4   Alkynes

Topic Test 5: Acidity of Alkynes and Alkylation of Acetylides

True/False

1. The acidity of a hydrogen attached to carbon increases with the amount of s character in the bond to that hydrogen.

2. Acetylides can be alkylated with a methyl halide or primary halide, but tertiary alkyl halides do not work in that reaction.

Multiple Choice

3. Which of the following is the most acidic?
   a. Ethane
   b. Ethene
   c. Ethyne
   d. Ammonia
   e. None of the above

4. Which of the following would be least likely to result from the alkylation of an acetylide?
   a. 2-Butyne
   b. 2,2,5,5-Tetramethyl-3-hexyne
   c. 4,4-Dimethyl-2-pentyne
   d. All of the above
   e. None of the above

Short Answer

5. Complete the following reaction scheme.

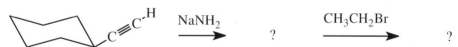

6. Show how one could prepare 3-hexyne from 1-butyne and any other needed reagents.

Topic Test 5: Answers

1. **True.** Because electrons in an s orbital are closer to the nucleus and more tightly held, one can predict that an sp-hybridized carbon is more electronegative than sp^2- or sp^3-hybridized carbons. Thus, an sp-hybridized carbon is better able to accommodate the negative charge on the conjugate base.

2. **True.** For the acetylide to attach onto the carbon of the alkyl halide, it must approach from the back side (i.e., opposite side from the halogen). If the alkyl halide is tertiary, the back side is too crowded and the approach is blocked.

3. **c.** Because the alkyne carbons have more s character in the orbitals that hold the hydrogens, we expect the alkyne hydrogens to be the most acidic of the three hydrocarbons mentioned. Clearly, ammonia must be less acidic than ethyne because

otherwise amide ion would not be a strong enough base to deprotonate ethyne to yield the acetylide.

4. **b.** Because this alkyne has a *tert*-butyl group on each of the alkyne carbons, it would not be possible to make it via the acetylide alkylation route. (Recall that the alkylation reaction is only successful with methyl and primary halides.) 2-Butyne could be prepared from the reaction of CH_3I and the acetylide of propyne. 4,4-Dimethyl-2-pentyne could be prepared from CH_3I and the acetylide of 3,3-dimethyl-1-butyne.

5. cyclohexyl-C≡C-H $\xrightarrow{NaNH_2}$ cyclohexyl-C≡C:$^\ominus$ $\xrightarrow{CH_3CH_2Br}$ cyclohexyl-C≡C-CH_2CH_3

6. $CH_3CH_2C{\equiv}CH \xrightarrow{NaNH_2} CH_3CH_2C{\equiv}C{:}^\ominus \xrightarrow{CH_3CH_2Br} CH_3CH_2C{\equiv}CCH_2CH_3$

TOPIC 6: ALKYNE CLEAVAGE REACTIONS

KEY POINTS

✓ *What reagents can be used to cleave alkynes?*

✓ *What are the products of alkyne cleavage reactions?*

✓ *How can a cleavage reaction be used to deduce the structure of an alkyne?*

When alkynes are treated with strong oxidizing agents like ozone or permanganate in aqueous acid, the triple bond is cleaved and the resulting fragments are carboxylic acids. If the cleavage is performed on a terminal alkyne, then one of the fragments will be carbonic acid, which decomposes to carbon dioxide and water. The reactions below summarize these transformations.

$$R-C{\equiv}C-R' \xrightarrow[\text{or } O_3, H_3O^\oplus]{KMnO_4, H_3O^\oplus} RCO_2H + HO_2CR'$$

$$R-C{\equiv}C-H \xrightarrow[\text{or } O_3, H_3O^\oplus]{KMnO_4, H_3O^\oplus} RCO_2H + HOCOH \rightarrow CO_2 + H_2O$$

Topic Test 6: Alkyne Cleavage Reactions

True/False

1. Alkynes react with ozone in aqueous acid to yield ketones and aldehydes.

2. Although treatment of 2-hexyne with potassium permanganate in aqueous acid yields two different carboxylic acids, those same conditions convert 3-hexyne into only a single product.

Multiple Choice

3. Which of the following conditions will *not* cleave alkynes?
   a. H_2O, $HgSO_4$, H_2SO_4
   b. O_3, H_3O^+
   c. $KMnO_4$, H_3O^+
   d. All of the above
   e. None of the above

4. Which compound below will yield carbon dioxide as one of its cleavage products when treated with ozone and in aqueous acid?
   a. 1-Hexyne
   b. 2-Hexyne
   c. 3-Hexyne
   d. All of the above
   e. None of the above

Short Answer

5. Provide an unambiguous structural formula for any missing organic product(s).

   [Cyclic structure with $CH_2C{\equiv}CCH_2$ group, reacted with $KMnO_4$ / H_3O^+]

6. Deduce the structure of the missing reactant in the reaction scheme below.

   [Reaction: ? $\xrightarrow{O_3 / H_3O^+}$ $(CH_3)_2C(CO_2H)_2$ + $(CH_3)_2CHCO_2H$ + CO_2]

Topic Test 6: Answers

1. **False.** The products of alkyne ozonolysis are carboxylic acids (except in the case of a terminal alkyne where the terminal alkyne carbon becomes carbon dioxide).

2. **True.** Any symmetrical alkyne will yield two equivalents of the same product upon cleavage, whereas an unsymmetrical alkyne will necessarily yield two different cleavage products. 2-Hexyne will be converted into a two-carbon carboxylic acid and a four-carbon carboxylic acid (CH_3CO_2H + $HO_2CCH_2CH_2CH_3$), and 3-hexyne will be converted into two equivalents of a three-carbon carboxylic acid.

3. **a.** These are the conditions for alkyne hydration mentioned in Topic 3 of this chapter. Reagents b and c will cleave alkynes.

4. **a.** Terminal alkynes yield carbonic acid that in turn decomposes to carbon dioxide and water in a favorable equilibrium.

5. $HO_2C(CH_2)_7CO_2H$

6. $HC{\equiv}C-C(CH_3)_2-C{\equiv}C-CH(CH_3)_2$

APPLICATION

Several manufactured pharmaceuticals contain the alkyne linkage along with other functional groups. The structures of some of these are shown below along with their generic names and uses.

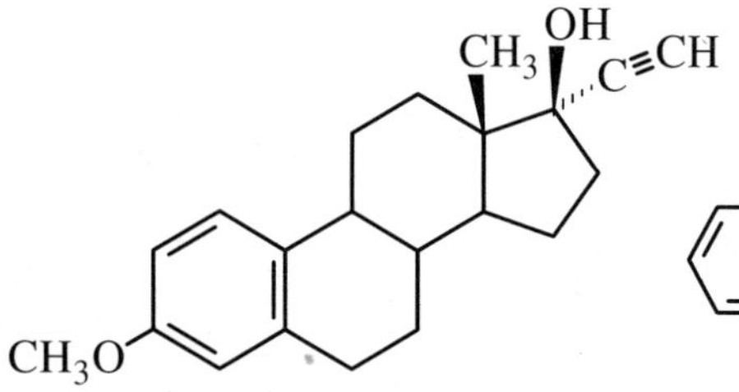

Mestranol
(oral contraceptive)

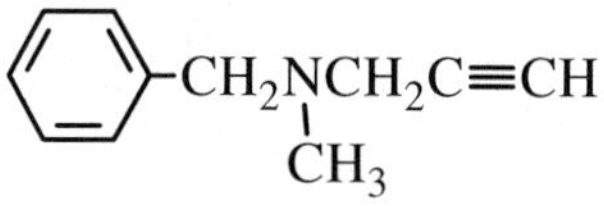

Pargyline
(antihypertensive)

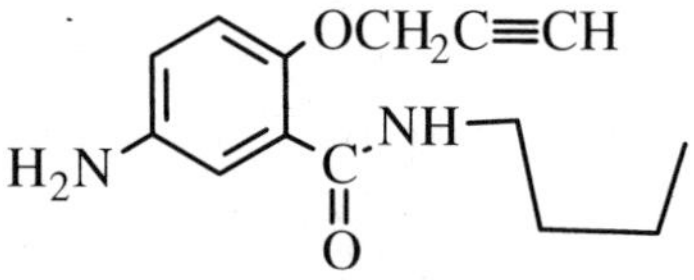

Parsalmide
(analgesic)

DEMONSTRATION PROBLEM

Show how one could synthesize the compound below from acetylene, alkyl halides with three or fewer carbons, and any needed inorganic reagents.

$$CH_3CH_2CH_2\overset{O}{\underset{\|}{C}}CH_3$$

Solution

A good strategy for successfully working a multistep synthesis problem is to work it backward, that is, identify the product by compound type and then ask yourself what reactions you know that will produce such compounds. This five-carbon ketone could not have been produced directly from three-carbon alkyl halides or acetylene using any reaction we have seen so far. Recall that hydration of alkynes gives enols that tautomerize to carbonyl compounds. Our ketone could have come from either of two enols. These are unstable intermediates that resulted from hydration of 1-pentyne or 2-pentyne as shown.

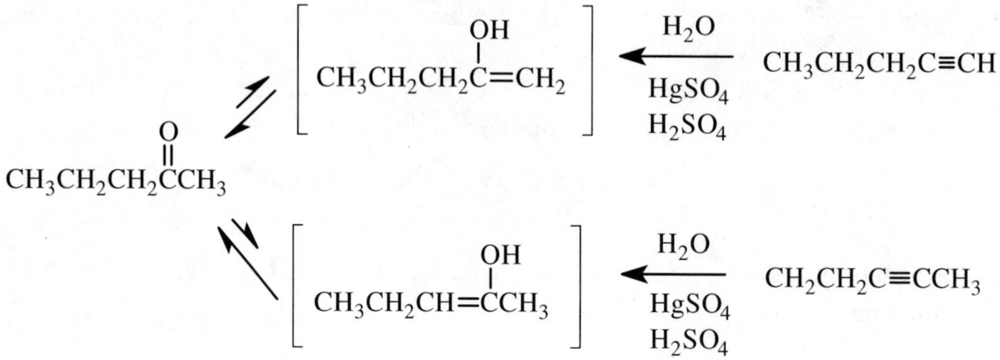

$$CH_3CH_2CH_2\overset{O}{\underset{\|}{C}}CH_3 \;\rightleftharpoons\; \left[CH_3CH_2CH_2\underset{OH}{\overset{|}{C}}=CH_2\right] \xleftarrow[H_2SO_4]{H_2O,\; HgSO_4} CH_3CH_2CH_2C\equiv CH$$

$$CH_3CH_2CH_2\overset{O}{\underset{\|}{C}}CH_3 \;\rightleftharpoons\; \left[CH_3CH_2CH=\underset{OH}{\overset{|}{C}}CH_3\right] \xleftarrow[H_2SO_4]{H_2O,\; HgSO_4} CH_3CH_2C\equiv CCH_3$$

Although hydration of either alkyne would give at least some of the desired ketone, the hydration of 1-pentyne is the preferable strategy because Markovnikov addition of water places the OH of the enol on carbon 2. The hydration of the unsymmetrical internal alkyne 2-pentyne would yield a mixture of two different enols that would tautomerize into two different ketones as shown.

$CH_2CH_2C\equiv CCH_3 \longrightarrow$ [enol with OH + enol with OH] ⇌ ketone + ketone

2-Pentyne could be made from acetylene and 1-bromopropane according to the reaction sequence from Topic 5. The organic starting materials allowed by the rules of the question are circled.

$CH_3CH_2CH_2C\equiv CH \xleftarrow{\boxed{CH_3CH_2CH_2Br}} :\overset{\ominus}{C}\equiv CH \xleftarrow{NaNH_2} \boxed{HC\equiv CH}$

Chapter Test

True/False

1. Alkenes do not react with H_2 and Lindlar's catalyst.
2. There are three different alkynes with the formula C_5H_8.
3. Alkynes can be cleaved to carboxylic acids with aqueous acid.
4. 2,2-Dimethyl-4-pentyne is a legitimate IUPAC name.
5. Lithium metal in liquid ammonia will convert a *cis* alkene into a *trans* alkene.

Multiple Choice

6. Which is true about alkynes?
   a. They have at least four atoms on the same axis.
   b. They have at least two pi bonds.
   c. They normally react with bromine to yield addition products.
   d. All of the above
   e. None of the above

7. Treatment of an unsymmetrical internal alkyne with borane followed by alkaline hydrogen peroxide would yield
   a. no reaction.
   b. two different stable enol products.
   c. two carboxylic acids.
   d. carbon dioxide.
   e. None of the above

8. Which of the following can be used to synthesize alkynes?
   a. Ozonolysis
   b. Tautomerization

c. Hydration
d. Hydrogenation
e. Dehydrohalogenation (elimination of HX)

9. Treatment of cyclooctyne with H_2O, H_2SO_4, $HgSO_4$ will produce
   a. cyclooctene.
   b. cycloactane.
   c.
   d.
   e. —OH

10. When 1-heptyne is treated with $KMnO_4$, H_3O^+, the resulting product is
    a. 1,3-heptadiene.
    b. $CH_3(CH_2)_4CO_2H + CO_2$.
    c. heptane.
    d. $HO(CH_2)_7OH$.
    e. None of the above

Short Answer and Essay

11. Small-ring cycloalkynes are notoriously unstable and cannot usually be isolated. For example, cyclopentyne has never been prepared and isolated. Suggest a possible reason for this.

12. Assume that the R groups below are unreactive and will not be effected by the reactions from this and previous chapters. Show how one could carry out the following conversion. More than one step may be required.

    $$RCH=CHR \rightarrow RC\equiv CR$$

    Provide unambiguous structural formulas for the organic products of the following reactions.

13. $(CH_3)_2CHCH_2C\equiv CCH_2CH=CH_2$ $\xrightarrow{\text{excess } H_2, \text{ Pd(C)}}$

14. $(CH_3)_2CHCH_2C\equiv CCH_2CH=CH_2$ $\xrightarrow{H_2, \text{ Lindlar's catalyst}}$

15. $(CH_3)_2CHCH_2C\equiv CCH_2CH=CH_2$ $\xrightarrow{\text{Li, ammonia}}$

16. Compound **A** has the formula C_6H_{10} and yields the products below when treated with hot acidic aqueous potassium permanganate. Deduce the structure of **A**.

    $$A \xrightarrow[H_3O^+, \text{ heat}]{KMnO_4} (CH_3)_3CCO_2H + CO_2$$

17. What reagents or conditions could one use to synthesize the ketone below from an alkyne and any needed inorganic reagents?

18–21. Complete the following reactions.

18. 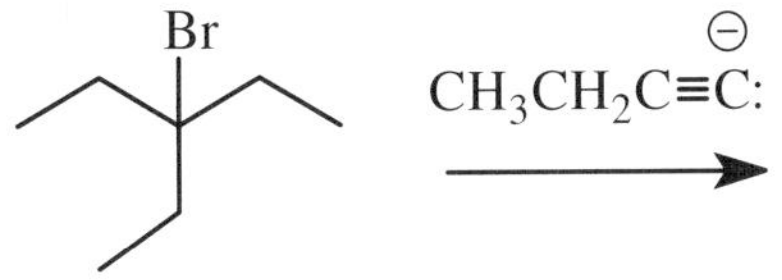 [structure: 3-bromo-3-methylpentane] + CH₃CH₂C≡C:⁻ →

19. 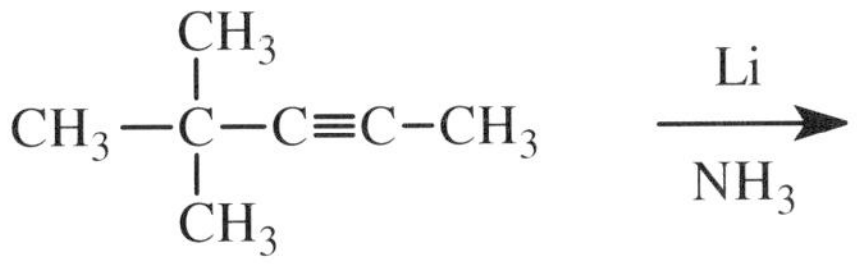 [structure: (CH₃)₃C–C≡C–CH₃] $\xrightarrow[\text{NH}_3]{\text{Li}}$

20. 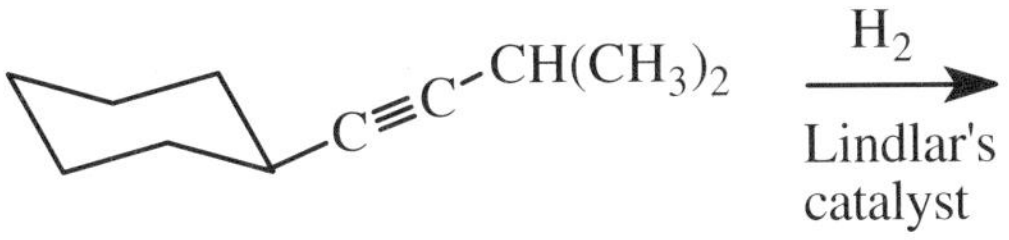 [structure: cyclohexyl–C≡C–CH(CH₃)₂] $\xrightarrow[\text{Lindlar's catalyst}]{\text{H}_2}$

21. 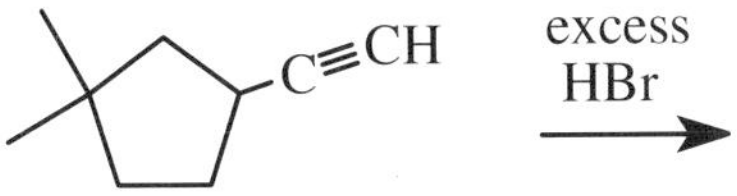 [structure: gem-dimethylcyclopentane with C≡CH substituent] $\xrightarrow{\text{excess HBr}}$

22. Show how one could synthesize 2-hexyne from propyne and any other needed reagents.

Chapter 4 Test: Answers

1. **True.** (Otherwise how would the reaction of an alkyne stop at the alkene stage?)

2. **True.** (1-pentyne, 2-pentyne, and 3-methyl-1-butyne).

3. **False.** (Ozonolysis or treatment with permanganate will work but not just H_2O/H_3O^+.)

4. **False.** The structure implied by this incorrect name would actually be called 4,4-dimethyl-1-pentyne.

5. **False.** These reagents will convert an alkyne into a *trans* alkene.

6. **d** 7. **e** 8. **e** 9. **c** 10. **b**

11. The internal angle of a pentagon is much smaller than the 180-degree bond angles required for alkynes.

12. Br_2/CCl_4 yields RCHBrCHBrR and then excess $NaNH_2$ gives the alkyne by a double dehydrobromination.

13. $(CH_3)_2CHCH_2CH_2CH_2CH_2CH_2CH_3$

14. $(CH_3)_2CHCH_2CH=CHCH_2CH=CH_2$ (*cis* around C_4 alkene)

15. $(CH_3)_2CHCH_2CH=CHCH_2CH=CH_2$ (*trans* around C_4 alkene)

16. $(CH_3)_3CC≡CH$

17. Hydration of 2,5-dimethyl-3-hexyne can be done with water in the presence of mercury (II) sulfate and sulfuric acid. Another strategy would be to treat the same alkyne with borane followed by alkaline hydrogen peroxide.

18. [structure] + CH₃CH₂C≡CH (elimination)

19. [structure] (trans)

20. [structure] (cis)

21. [structure with Br, Br]

22. CH₃C≡CH —NaNH₂→ CH₃C≡C:⁻ —BrCH₂CH₂CH₃→ CH₃C≡CCH₂CH₂CH₃

Check Your Performance

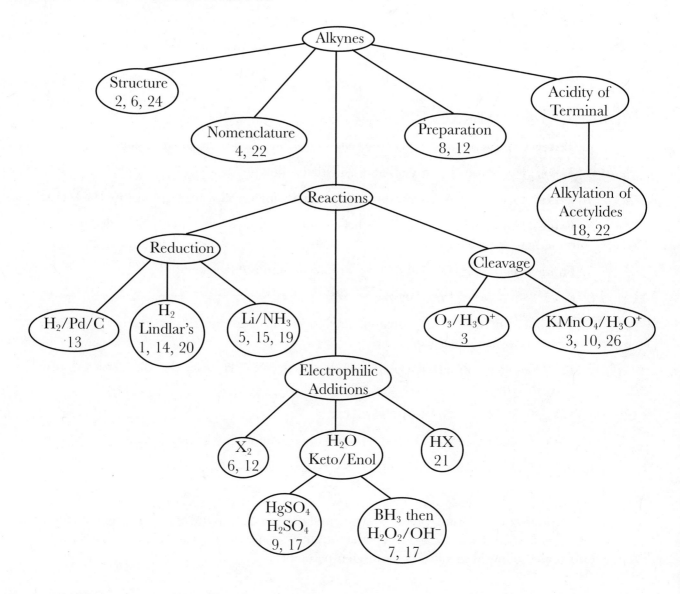

CHAPTER 5

Stereochemistry

The scent of caraway seeds is quite different from that of spearmint leaves, yet both aromas are caused by the natural product carvone. The carvone from these two sources differs in an important way that we explore in this chapter.

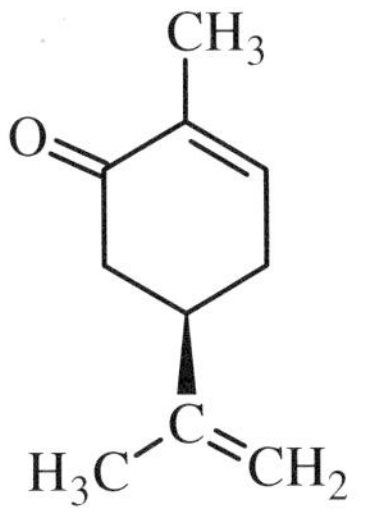

(R)-Carvone
spearmint oil

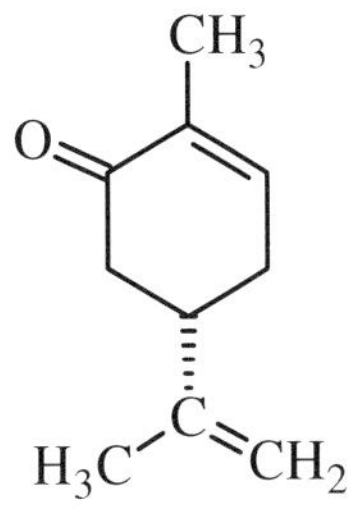

(S)-Carvone
caraway seed oil

ESSENTIAL BACKGROUND

- Drawing structural formulas (Chapter1)
- Molecular shapes (Chapter 1)
- Definition of isomers (Chapter 2)
- *Cis* and *trans* (Chapter 2)
- Cahn-Ingold-Prelog (CIP) prioritizing of atoms and groups (Chapter 3)

TOPIC 1: ISOMER CLASSIFICATION

KEY POINTS

✓ *What are constitutional isomers?*

✓ *What are stereoisomers?*

Recall from Chapter 2 that molecules with the same formula but different structures are called isomers. Some isomers differ by the connectivity of atoms, that is, the linkage between atoms is not the same. Such isomers are called **constitutional isomers**. The difference can be clear and obvious such as cyclopentanol and 2-pentanone shown below.

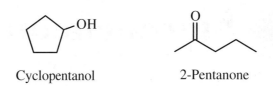

Cyclopentanol   2-Pentanone

Both have the formula C₅H₁₀O, yet one is a cyclic alcohol whereas the other is a ketone. The difference between constitutional isomers can be subtle. Consider *trans*-3-octene and *trans*-4-octene. Both are unbranched internal *trans* alkenes with the formula C₈H₁₆ but the double bond is centered in one compound and off center in the other. Because the order of linkage is different they are constitutional isomers.

trans-3-Octene   *trans*-4-Octene

Another major subclass of isomers are **stereoisomers**. These have the same formula and order of linkage but differ in how the atoms are oriented in three-dimensional space. Consider *cis*- and *trans*-1,2-dimethylcyclopropane. Each is a three-membered ring hydrocarbon with methyl groups on carbons 1 and 2, yet the *cis* isomer will always be different from the *trans* with respect to the orientation of the methyl groups. Interconversion of these stereoisomers would require breaking bonds. A similar difference exists for *cis*- and *trans*-2-butene.

cis-1,2-Dimethylcyclopropane   *trans*-1,2-Dimethylcyclopropane   *cis*-2-Butene   *trans*-2-Butene

Stereoisomers can differ in even less obvious ways. Consider the pictures of 2-hydroxypropanoic acid shown below. These are quite similar, having the same formula and linkage, yet they are not identical. They are mirror image isomers and cannot be interconverted without breaking bonds. We discuss this more later, but for now just note that they are stereoisomers.

(*S*)-2-Hydroxypropanoic acid   (*R*)-2-Hydroxypropanoic acid

Topic Test 1: Isomer Classification

True/False

1. 1-Butyne and 2-butyne are stereoisomers.
2. Hexane and cyclohexane are stereoisomers.
3. Constitutional isomers may contain different functional groups.

Multiple Choice

4. Which pair of compounds below are stereoisomers?
    a. 1-Pentene and *trans*-2-pentene
    b. *cis*-2-pentene and cyclopentane
    c. *cis*-2-pentene and *trans*-2-pentene
    d. All of the above
    e. None of the above

5. Which pair of compounds below are constitutional isomers?
    a. Cyclohexane and *trans*-3-hexene
    b. 1,3-Butadiene and 2-butyne
    c. Methylcyclobutane and 1-pentene
    d. All of the above
    e. None of the above

Short Answer

6. Can stereoisomers contain different functional groups? Explain.

Topic Test 1: Answers

1. **False.** The triple bond of 1-butyne is between carbons 1 and 2, whereas that of 2-butyne is between carbons 2 and 3. These are constitutional isomers because they have the same formula but different order of linkage.

2. **False.** These two have different formulas and are therefore not any kind of isomers.

3. **True.** Constitutional isomers must have the same formula but differ by the connectivity of the atoms (e.g., cyclopentanol and 2-pentanone mentioned in this section).

4. **c.** *cis*-2-pentene and *trans*-2-pentene are stereoisomers because they are each a four-carbon chain with an alkene linkage between carbons 2 and 3. Each has the formula C_5H_{10} and differs from the other only in how the atoms are directed in space. Pairs a and b are constitutional isomers.

5. **d.** The pairs of compounds named all have the same formulas but different connectivity.

6. **No.** Stereoisomers will necessarily contain the same functional groups because they differ only by the orientation of the atoms in space but will have the same connectivity and linkage.

TOPIC 2: CHIRALITY

KEY POINTS

✓ *What is meant by the term chiral?*

✓ *How can one identify chiral molecules?*

✓ *What is a stereogenic center?*

✓ *What is a* meso *compound?*

Chiral is an adjective used to describe any object that will not superimpose on its mirror image. Enantiomers are chiral in that they have a relationship similar to your right and left hand. They are alike in many ways yet exist as right-handed and left-handed versions. The opposite of chiral is **achiral** (i.e., not chiral). Because of their symmetry, achiral objects are superimposable on (identical to) their mirror image. To determine if a molecule is chiral, one can use the test of superimposition on a mirror image. Many (though not all) chiral molecules contain at least one tetrahedral center that bears four different groups. Such centers are sometimes called **stereogenic centers** (or, more ambiguously, asymmetric centers or chiral centers, but these terms can be misleading). Another way to test for chirality is to look for a plane of symmetry. The presence of a symmetry plane indicates that the object will superimpose on its mirror image and therefore is achiral. Bromoiodomethane is achiral and has a plane of symmetry. Bromofluoroiodomethane is chiral.

Plane of symmetry, Achiral   Chiral

Some molecules contain two or more stereogenic centers, yet overall the molecules are not chiral. These are called **meso** compounds. They can be detected by inspection in that they contain a plane of symmetry. One isomer of 2,3-dibromobutane illustrates this concept.

meso-2,3-Dibromobutane
plane of symmetry
not chiral

($2S,3S$)-2,3-Dibromobutane ($2R,3R$)-2,3-Dibromobutane
Pair of enantiomers

Topic Test 2: Chirality

True/False

1. A shoe is chiral.

2. Butane contains two stereogenic centers.

Multiple Choice

3. Which of the following is (are) chiral?
   a. A standard carpentry nail
   b. A standard machine screw

c. An undecorated coffee mug
d. All of the above
e. None of the above

4. How many stereogenic centers are contained in 3-methylhexane?
   a. one
   b. two
   c. three
   d. four
   e. none

Short Answer

5. Redraw the structural formula below and place an asterisk beside each stereogenic center.

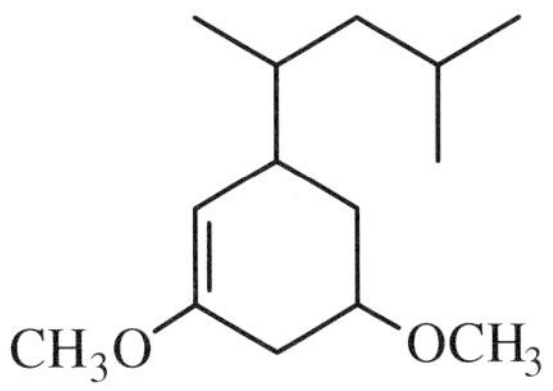

6. Provide an unambiguous structural formula for a chiral alcohol having the formula $C_4H_{10}O$.

Topic Test 2: Answers

1. **True.** It usually matters whether you put a shoe on your right foot or your left foot, and it will clearly fit better on one foot than it will on the other. A shoe is constructed to fit over a chiral foot and is nonsuperimposable on its mirror image.

2. **False.** Each of the four carbons of butane bears either two or three hydrogens. Stereogenic centers have four different ligands around a tetrahedron.

3. **b.** The threads of the screw have an asymmetry that pulls it into a hole when turned in one direction and lifts it out of the hole when turned in the other direction. A screw is not superimposable on its mirror image and is therefore chiral.

4. **a**

5.

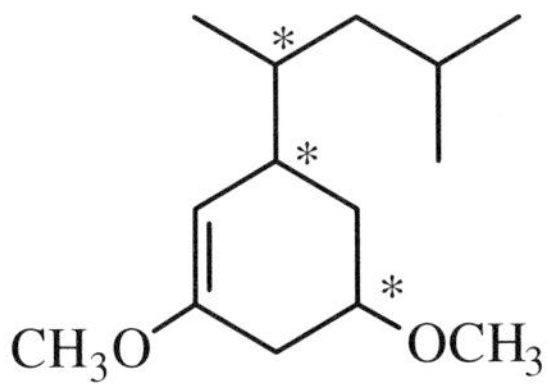

6.

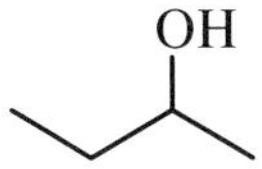

TOPIC 3: CLASSIFICATION AND NUMBER OF STEREOISOMERS

KEY POINTS

✓ *What are enantiomers?*

✓ *What are diastereomers?*

✓ *How many stereoisomers are possible for a molecule with n stereogenic centers?*

Nonsuperimposable mirror image isomers are called **enantiomers**. Stereoisomers that are not enantiomers are called **diastereomers** (i.e., diastereomers are stereoisomers that are not mirror images). There are several subcategories of diastereomers, but for now it will be simpler to recognize two major classes of stereoisomers: enantiomers (mirror image isomers) and diastereomers (all other stereoisomers). Enantiomers often contain stereogenic centers as illustrated below. Note that there is no orientation of either structure that will make it exactly like the other.

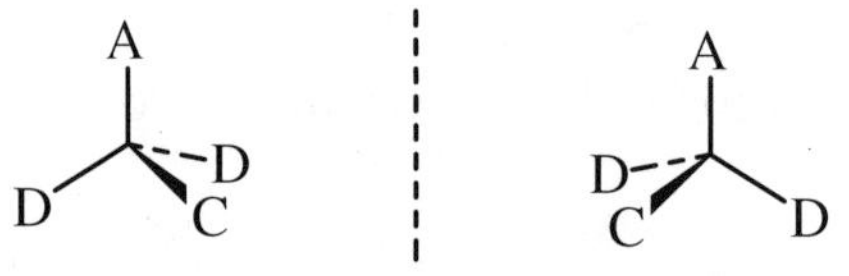

It is also possible to have larger and more complex pairs of enantiomers containing multiple stereogenic centers, but they will always be nonsuperimposable mirror images.

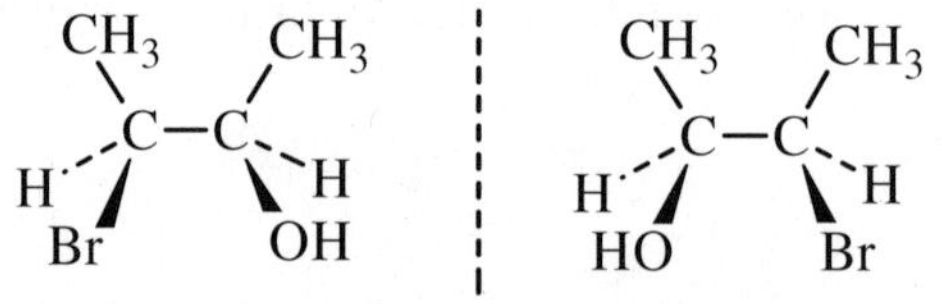

Diastereomers can be *cis/trans* isomers as shown below for substituted cyclic compounds or alkenes.

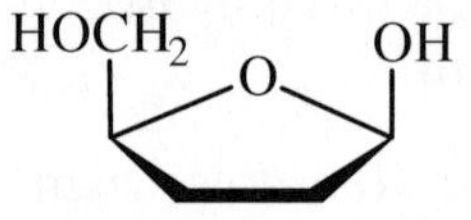

Cis 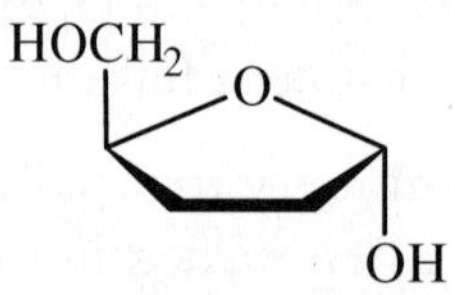   Trans    *cis*-1,3-Pentadiene 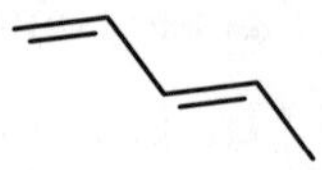   *trans*-1,3-Pentadiene

A subtler form of diastereomers exists when there is more than one stereogenic center in a molecule. The naming of these is discussed later, but for now consider the two isomers of 2,3-dibromobutane shown below. Although there is free rotation around the sigma bonds for each compound, there is no conformation that will align the atoms in the exact same orientation as the other isomer. Because these stereoisomers are not mirror images and therefore cannot be enantiomers, they must be diastereomers.

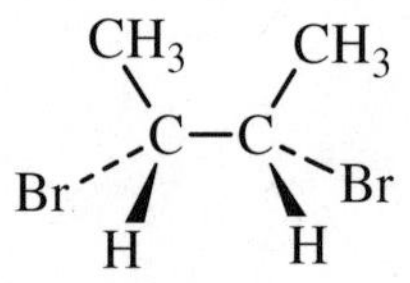

(*meso*)-2,3-Dibromobutane 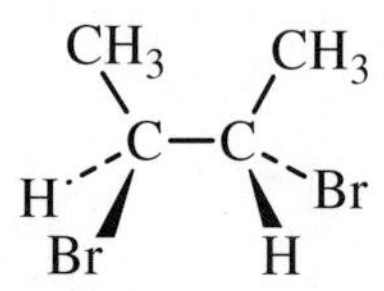   (2*S*,3*S*)-2,3-Dibromobutane

A molecule with n stereogenic centers can have up to 2^n possible stereoisomers (i.e., 2 raised to the nth power). A linkage containing 1, 2, 3, 4, or 5 stereogenic centers can have 2, 4, 8, 16, or 32 stereoisomers, respectively. There may be fewer than 2^n stereoisomers in some cases if *meso* compounds exist.

Topic Test 3: Classification and Number of Stereoisomers

True/False

1. Stereoisomers that are not mirror images are called enantiomers.
2. *cis*- and *trans*-1,2-dimethylcyclobutane are diastereomers.

Multiple Choice

For each pair of structural formulas shown in problems 3 through 5, specify whether they are
a. enantiomers
b. diastereomers
c. the same
d. None of the above
e. All of the above

3.

4.

5.

Short Answer

6. Use wedges and dashed lines to draw three-dimensional representations of the three possible stereoisomers that are 2,3-difluorobutane. Specify the relationship between each possible pair of isomers as enantiomers or diastereomers.

7. What is the maximum number of stereoisomers possible for 1,2,3,4-tetrabromopentane?

Topic Test 3: Answers

1. **False.** Enantiomers must be mirror images. Stereoisomers that are not enantiomers are diastereomers.

2. **True.** They are not mirror images (and therefore cannot possibly be enantiomers), but they are stereoisomers.

3. **c.** Both pictures represent 2-methylpentane, which has no stereogenic centers and is not chiral. It also has no rings or double bonds to allow *cis/trans* isomerization.

4. **a.** These are nonsuperimposable mirror images. At first glance they may not appear to be mirror images, yet if one imagines lifting one of the structures out of the page and holding it "in front" of the other with the "mirror" between the two, the relationship is easier to see.

5. **b.** The only stereogenic center in these compounds has the same configuration, but one of the structures is *cis* and the other is *trans* around the alkene linkage.

6.

      A   B   C

Enantiomers: **B** and **C**. Diastereomers: **A** and **B**; **A** and **C**

7. **Eight.** This structure has three stereogenic centers at carbons 2, 3, and 4. The maximum number of stereoisomers one predicts is $2^3 = 8$.

TOPIC 4: ASSIGNING CONFIGURATION AROUND STEREOGENIC CENTERS

KEY POINTS

✓ *How is configuration around a stereogenic center indicated?*

✓ *What do the prefixes R and S mean?*

A stereogenic center can have either of two possible configurations. These are assigned and named according to the relative positions of the various ligands. Prioritize the four ligands around the stereogenic center according to CIP conventions that were used earlier to designate *E* and *Z* alkenes. Imagine the lowest priority ligand away from you and note the direction your eye travels as you scan through the three highest priority ligands in order. If clockwise, the designator *R* (*rectus*, right) is used. If counterclockwise, the designator *S* (*sinister*, left) is used.

Sometimes the inspection and assignment of *R* or *S* is made more difficult in that the lowest priority ligand is not oriented away from the viewer. In those cases one must imagine what the stereogenic center would look like if the viewer were appropriately positioned. (It is usually easier and more reliable to move yourself rather than trying to redraw the picture and risking unintended isomerization.) Below are some examples of stereogenic centers with indications of the proposed viewer's line of sight.

R (visualize view from the top)   S (visualize view from behind the page)

The prefixes *R* or *S* are part of the complete name of a molecule that contains stereogenic centers. These designators are normally italicized and enclosed within parentheses. In the case of multiple stereogenic centers, numbers are used to specify which ones are *R* and which are *S*.

Topic Test 4: Assigning Configuration Around Stereogenic Centers

True/False

1. The absolute configuration around a stereogenic center is assigned by placing the smallest ligand in the "back" and scanning through the remaining ligands largest to smallest.

2. The mirror image of an *R* stereogenic center will be an *S* stereogenic center.

Multiple Choice

3. The stereochemistry of the compound shown below is

    a. 2*R*, 3*R*
    b. 2*R*, 3*S*
    c. 2*S*, 3*S*
    d. 2*S*, 3*R*
    e. None of the above

4. The term that best describes the stereochemistry of the compound below is

    a. *trans*
    b. *cis*

Topic 4: Assigning Configuration Around Stereogenic Centers   97

c. *R*
   d. *S*
   e. achiral

Short Answer

5. Provide a complete name (including stereochemical designator) for the compound below.

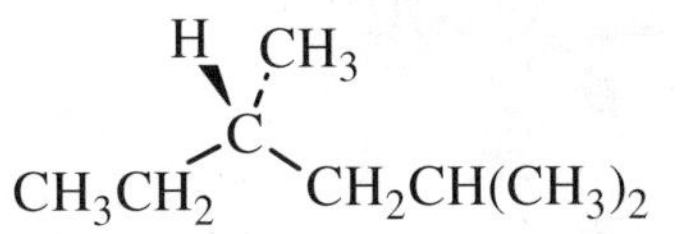

$$\underset{CH_3CH_2}{\overset{HCH_3}{\diagdown\,\diagup}}\underset{}{\overset{C}{}}\underset{CH_2CH(CH_3)_2}{}$$

6. Draw a clear three-dimensional representation of (*R*)-2-bromobutane.

Topic Test 4: Answers

1. **False.** The *R* and *S* configurations come from the CIP priorities that are based on atomic numbers (not size). Recall, for example, that fluorine (atomic number 9) is higher priority than a t-butyl group, which is larger but attached at carbon (atomic number 6).

2. **True.** Enantiomers have the opposite configurations around their stereogenic centers.

3. **a**

4. **d.** Imagine what the stereogenic center would look like if viewed from slightly below and inside the ring. The low-priority methyl will appear behind the page away from the viewer.

$$OCH_2 > OH > CH_2$$

5. (*S*)-2,4-dimethylhexane. There is one stereogenic center along the parent six-carbon chain. Its configuration is shown three dimensionally, but the lowest priority ligand is toward the viewer. Tracing through the three highest priority ligands in order (isobutyl > ethyl > methyl) appears to lead the eye in the clockwise direction; however, it would appear to be the counterclockwise direction if viewed from the other side with the hydrogen away from the viewer.

6. There are many possible correct representations. Several are shown here.

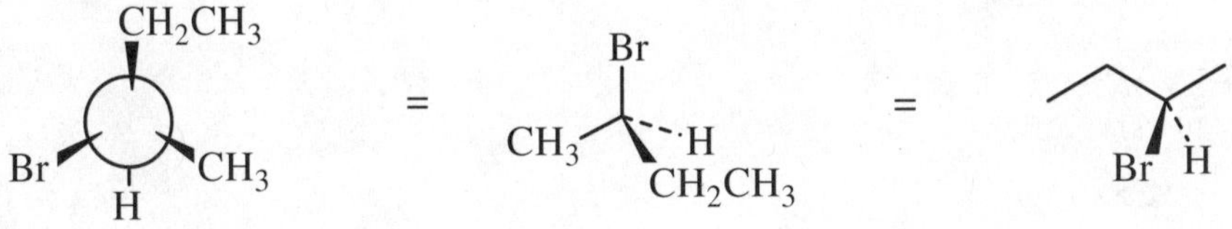

TOPIC 5: FISCHER PROJECTIONS

KEY POINTS

✓ *What are Fischer projections?*

✓ *What are the conventions for drawing and interpreting Fischer projections?*

A tetrahedral carbon can be viewed from many different orientations as shown below. Pictures **I** and **II** show the central atom and two of the four bonds in the plane of the page. A different picture of the same structure is given by **III**, where only the central atom is in the plane of the page, whereas groups 3 and 4 come out toward the viewer and groups 1 and 2 are behind the plane of the page. From this perspective it appears as though 3-C-4 are in a straight horizontal line and 2-C-1 appears to be in a straight vertical line. (This is, of course, not the case but only appears as such due to perspective.) In a notation called **Fischer projections**, the wedges and dashed lines of picture **III** that were used to imply in front of and behind the plane of the page, respectively, are not used. These are instead represented by solid lines, but we are to agree that, by convention, bonds toward the viewer are horizontal and bonds away from the viewer are vertical.

Although Fischer projections are faster and easier to draw, the opportunity for error increases when they are used. Do not rotate Fischer projections in the page 90 degrees or any odd multiple of 90 degrees when trying to make stereochemical predictions. Such movements result in accidental unintended inversion of all stereogenic centers. Fischer projections are especially useful for molecules such as carbohydrates that have multiple stereogenic centers.

Note that often Fischer projections are oriented such that the lowest priority ligand around a stereogenic center is attached to the molecule by a horizontal line, implying that it is pointing out of the page toward the viewer. This may make assigning the R/S configuration slightly more difficult.

Topic Test 5: Fischer Projections

True/False

1. The vertical lines of a Fischer projection indicate bonds that are pointed away from the viewer.

2. Rotating a Fischer projection by 90 degrees will invert the configuration of any stereogenic center.

Multiple Choice

3. Which of the following is a Fischer projection of the molecule pictured?

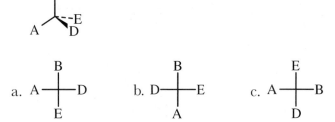

a. A—|—D with B on top, E on bottom
b. D—|—E with B on top, A on bottom
c. A—|—B with E on top, D on bottom

d. All of the above
e. None of the above

4. The term that best describes the relationship represented by the Fischer projections below is

$$\begin{array}{c} CO_2H \\ H-\!\!\!|\!\!\!-OH \\ CH_2OH \end{array} \qquad \begin{array}{c} CO_2H \\ HO-\!\!\!|\!\!\!-H \\ CH_2OH \end{array}$$

a. same
b. enantiomers
c. diastereomers
d. constitutional isomers
e. None of the above

Short Answer

5. Provide a complete name (including stereochemistry) for the molecule pictured below.

$$\begin{array}{c} CH_2CH_3 \\ H-\!\!\!|\!\!\!-CH_3 \\ CH=CH_2 \end{array}$$

6. Draw a Fischer projection of a three-carbon aldehyde that bears a hydroxy group on C2 and has the *R* configuration. Place the carbons along the apparent vertical line of the picture with the aldehyde at the top.

Topic Test 5: Answers

1. **True** by definition.

2. **True.** Because vertical and horizontal lines have spatial meaning in a Fischer projection, a 90-degree rotation changes the implied molecule to the opposite configuration.

3. **d.** Let the priorities be alphabetical (A > B > D > E) and note that all three Fischer projections are of the *R* enantiomer just viewed from different perspectives.

4. **b.** Both are representations of a three-carbon carboxylic acid having hydroxy groups on carbons 2 and 3, but the two pictures represent opposite absolute configurations around the stereogenic center.

5. (*S*)-3-methyl-1-pentene

6.

$$\begin{array}{c} H\diagdown \!\!\!\!{}_{C}\!\!\diagup\!\!{}^{O} \\ H\!\!-\!\!\!\!\!\!\!\!\!\!\!\!-\!\!\!\!\!\!-OH \\ CH_3 \end{array}$$

TOPIC 6: POLARIMETRY AND ASSOCIATED TERMINOLOGY

KEY POINTS

✓ *What is plane-polarized light?*

✓ *What is a polarimeter?*

✓ *What is meant by the term "optically active"?*

✓ *What kinds of solutions are optically active?*

✓ *How are optical rotations standardized and expressed?*

Normal light can be thought of as waves occupying many planes that are all at right angles to the direction of propagation. If a beam of normal light is directed toward a polarizing filter, nearly all the waves are blocked from passing through and the light that emerges is polarized into a single plane. This light is called **plane-polarized**. The plane can be rotated if it passes through an asymmetrical environment. A device used for creating plane-polarized light and measuring its rotation is called a **polarimeter**.

When plane-polarized light is passed through some liquids, the orientation of the plane changes. Solutions that rotate plane-polarized light are called **optically active**. A requirement for optical activity is the presence of an excess of one enantiomer. Enantiomers rotate light in opposite directions by equal amounts. Compounds that rotate plane-polarized light to the right or left are called **dextrorotatory** or **levorotatory**, respectively. Generally, solutions of chiral compounds are optically active.

An equal concentration of enantiomers (a **racemic mixture** or **racemate**) is not optically active because half the molecules rotate the plane in one direction and the other half of the molecules reverse that effect so that the resulting beam of plane-polarized light will experience no net change. Likewise, compounds that are not chiral are not optically active. *Meso* compounds and symmetrical compounds do not exhibit optical activity. The direction and magnitude of optical rotation is a characteristic for a given chiral compound. These values can be used for characterization and identification of a compound provided the conventions for measuring and reporting them are established. The **specific rotation** of a compound is given by the expression

$$[\alpha]_D = \frac{\alpha}{(l \times C)}$$

where $[\alpha]_D$ is the specific rotation at 589 nm (commonly called the sodium D line); α is the measured rotation, which is an experimentally generated value expressed in degrees; l is the path length through which light passes measured in decimeters; and C is the concentration expressed in (g/mL).

Topic Test 6: Polarimetry and Associated Terminology

True/False

1. The specific rotations of two enantiomers will be equal in magnitude but opposite in sign.
2. *Meso* compounds are optically active.

Multiple Choice

3. Which of the following is optically active?
   a. A racemic mixture of 2-bromobutane
   b. 2-Bromopropane
   c. (*S*)-2-Bromobutane
   d. All of the above
   e. None of the above

4. What would be the specific rotation of a compound that gave an observed rotation of +27.0 degrees when a solution of 0.900 g/mL was measured through a 1.00-decimeter path length?
   a. +24.3
   b. −24.3
   c. +30.0
   d. −30.0
   e. None of the above

Short Answer

5. Propose a structure for an optically active aldehyde with the formula $C_5H_{10}O$.
6. A device that measures rotations of plane-polarized light is called a(n) _____.
7. Compounds that rotate plane-polarized light to the left are called _____.

Topic Test 6: Answers

1. **True**
2. **False.** Although they contain two or more stereogenic centers, *meso* compounds are not chiral and therefore do not rotate plane-polarized light.
3. **c.** (*S*)-2-bromobutane is the only choice that contains an excess of one enantiomer. The racemic mixture of a contains an equal mixture of enantiomers, resulting in no net rotation, and the compound b is not chiral.

4. **c.** +30.0 = +27.0/(1.00 × 0.900)

5.  Either enantiomer will be optically active.

6. polarimeter

7. levorotatory

TOPIC 7: PROPERTIES OF STEREOISOMERS AND RESOLUTION OF RACEMIC MIXTURES

KEY POINTS

✓ *What are the relative chemical and physical properties of enantiomers?*

✓ *What are the relative chemical and physical properties of diastereomers?*

✓ *What is the process of separating a racemic mixture into enantiomers called?*

✓ *How are enantiomers separated?*

✓ *How can one predict if the product of a reaction will be optically active?*

Enantiomers generally have identical chemical and physical properties except for the direction of their specific rotations or when they are placed in an unsymmetrical environment. For example, two different enantiomers will likely interact differently with some other chiral compound. Sometimes two enantiomers will have slightly different aromas because olfactory receptors are themselves composed of chiral sites. The situation is roughly like a (chiral) right hand sorting through a mixture of both right and left (each chiral) gloves. The right gloves will interact differently with the right hand than will the left gloves. Enantiomers behave identically under most circumstances and in most symmetrical environments. They will have the same boiling point, melting point, and solubility in achiral solvents. For this reason, mixtures of enantiomers require a special strategy to separate the components.

Diastereomers usually have similar but not identical properties. Obviously some diastereomers are close structural analogues, so we would not expect their properties to differ drastically, yet they do have differences that can be used to separate them from one another. Two diastereomers will likely have different boiling points, melting points, and solubilities in achiral solvents.

The process of separating a racemic mixture into its component enantiomers is called **resolution**. The resolution of a racemic mixture is done according to a strategy that exploits the differing properties of diastereomers. The process generally involves reacting the racemic mixture of **A** with a single enantiomer of some other chiral compound **B**. The resulting products will be diastereomers that can be separated according to differences in their physical properties. Once isolated, these diastereomers can be treated to reverse the chemistry of the first reaction to yield individual enantiomers as shown schematically below.

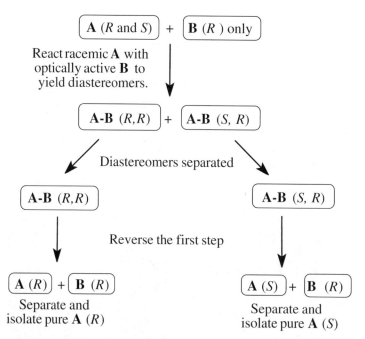

Normally, optically active products are obtained from reactions of optically active starting materials. Even reactions that give rise to new stereogenic centers will yield racemic mixtures under normal conditions, as shown below for the addition of HBr to 1-butene. An equal probability of the bromide attacking either face of the planar carbocation intermediate requires an equal mixture of enantiomers for the product.

Topic Test 7: Properties of Stereoisomers and Resolution of Racemic Mixtures

True/False

1. Diastereomers have identical physical and chemical properties.
2. The reaction $CH_3CH=CH_2 + Br_2 \rightarrow CH_3CHBrCH_2Br$ will yield an optically active product.

Multiple Choice

3. Which is true about enantiomers?
   a. They have identical boiling points.

104   Chapter 5   Stereochemistry

b. They have identical melting points.
c. They have identical solubility in achiral solvents.
d. All of the above
e. None of the above

4. Which is true about diastereomers?
   a. They have identical boiling points.
   b. They have identical melting points.
   c. They have identical solubility in achiral solvents.
   d. All of the above
   e. None of the above

Short Answer

5. The separation of enantiomers from a racemic mixture is called _____.

6. Briefly describe in words and/or schematically how a racemic mixture is separated into enantiomers.

Topic Test 7: Answers

1. **False.** Diastereomers have different physical and chemical properties. These properties are often similar but are not identical.

2. **False.** Although the product is chiral, both enantiomers are present in equal amounts, and therefore it is not optically active.

3. **d**

4. **e**

5. resolution (of a racemic mixture)

6. The racemic mixture is treated with a single enantiomer of another chiral compound. The resulting product mixture is made up of diastereomers that are separated through their differing properties and then converted back into the original enantiomers. (A schematic example is found for racemic **A** treated by optically active **B** in the topic text above.)

APPLICATION

Ibuprofen is a popular analgesic marketed under the names Advil, Nuprin, and Motrin. Its structure is shown below. Note the stereogenic center indicated by the asterisk. As is common for chiral compounds, the two enantiomers of ibuprofen differ in their physiological effect. The S enantiomer has been found to act more than twice as fast as the R form. Most commercial ibuprofen is still sold as a racemic mixture due to the expense required to resolve it.

$(CH_3)_2CHCH_2$—⟨phenyl⟩—$\overset{*}{C}H(CH_3)CO_2H$ Ibuprofen

DEMONSTRATION PROBLEM

Draw Fischer projections for all the possible stereoisomers of the commpound shown below. Label each stereogenic center as *R* or *S*. Which pairs of isomers are enantiomers? Which pairs are diastereomers?

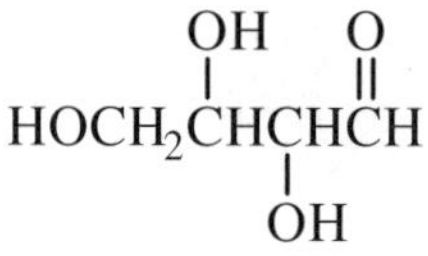

$$HOCH_2CHCHCH$$
with OH on first CH, OH on second CH, and =O on terminal CH (aldehyde)

Solution

Although there are many correct ways to draw the Fischer projections, it is convenient to proceed methodically using a common convention wherein the carbons of the chain are shown vertically and the aldehyde is at the top. There are two stereogenic centers so we predict a maximum of $2^2 = 4$ stereoisomers. Once they have been drawn, most of the *R/S* configurations are relatively easy to assess after the first one is determined. Say, for example, we work through the top stereogenic center in isomer A. (Note that the hydrogen is coming toward us so the eye's motion is reversed from what it would be if we were on the other side.) The priorities of the groups are OH > CH=O > bottom 2 carbons. The eye travels counterclockwise tracing through this sequence so the reverse would be clockwise and we conclude *R*. The top stereogenic centers in the other three isomers are now easily assigned as being the same as the one we just did (*R*) or the opposite (*S*). A similar process will give the configurations or the bottom four stereogenic centers. The end carbons are not stereogenic, and, therefore, it doesn't matter how they are facing in our pictures.

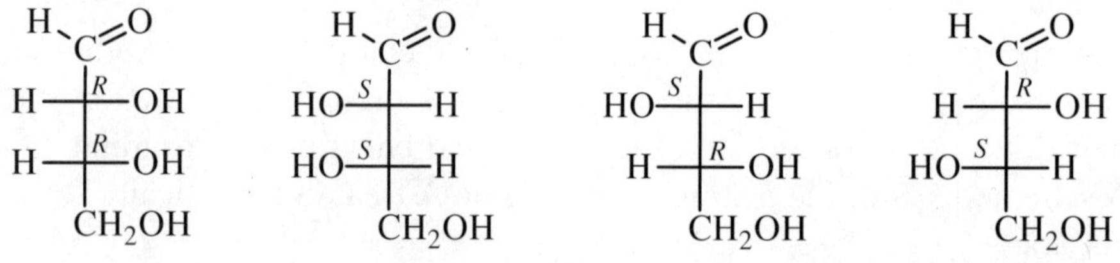

```
   H    C=O           H    C=O           H    C=O           H    C=O
   H─┼R─OH           HO─┼S─H            HO─┼S─H            H─┼R─OH
   H─┼R─OH           HO─┼S─H            H─┼R─OH            HO─┼S─H
     CH₂OH              CH₂OH              CH₂OH              CH₂OH

       A                  B                  C                  D
```

Enantiomers and diastereomers are selected by recognizing that enantiomers have opposite configurations throughout. All other stereoisomers are diastereomers. Pairs of enantiomers are A and B, C and D. Pairs of diastereomers are A and C, A and D, B and C, B and D.

Chapter Test

True/False

1. *Meso* compounds are chiral.
2. If the $[\alpha]_D$ of a given compound is -14.1, then the $[\alpha]_D$ for the enantiomer of that compound will be $+14.1$.

3. Stereoisomers will have the same functional groups.

4. A compound with a plane of symmetry is optically active.

Multiple Choice

5–17. Select the term from the list of choices that best describes the relationship for each pair shown below.
   a. Same
   b. Constitutional isomers
   c. Enantiomers
   d. Diastereomers
   e. None of the above

5.

6. Butane   Cyclobutane

7.

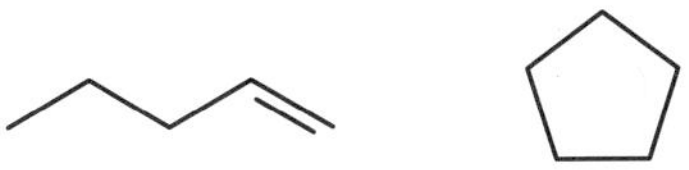

8.

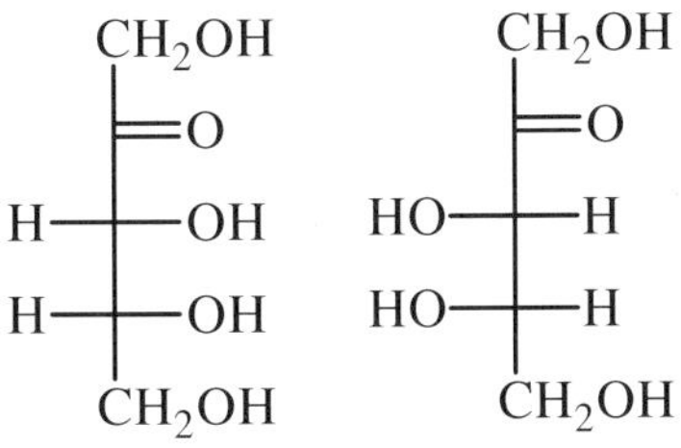

9.

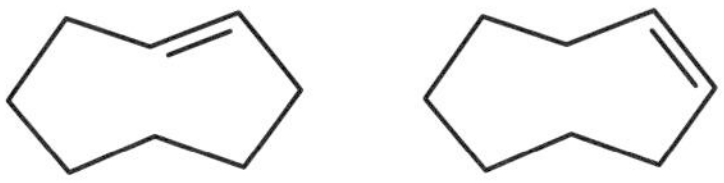

10.

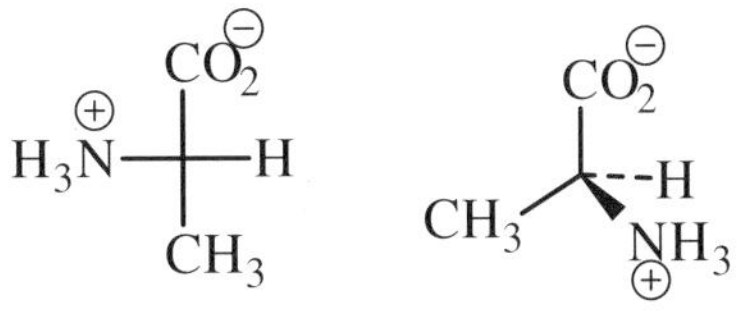

11.

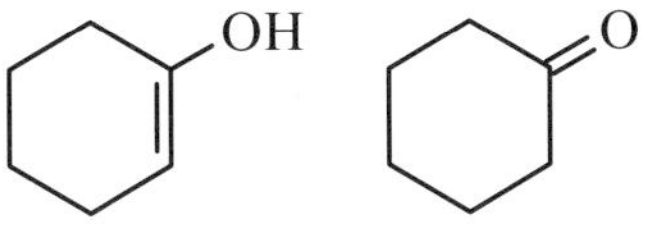

12.

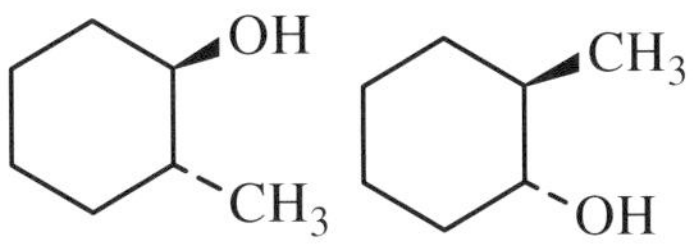

13. (R)-3-Bromo-1-butene   (S)-3-Bromo-1-butene

14.

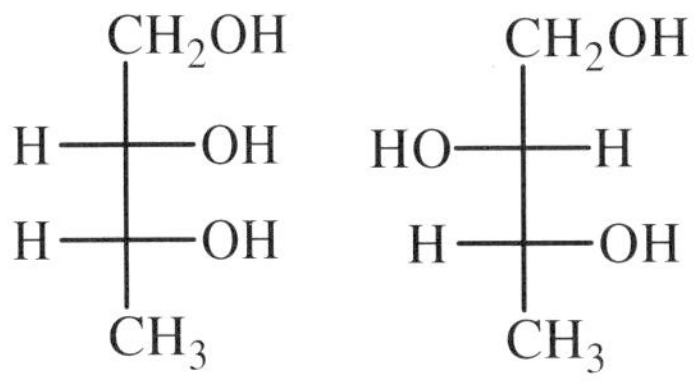

15.

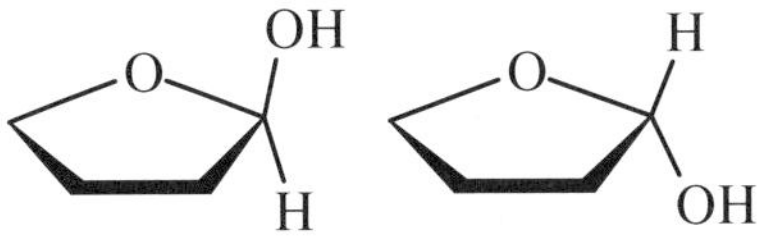

16.

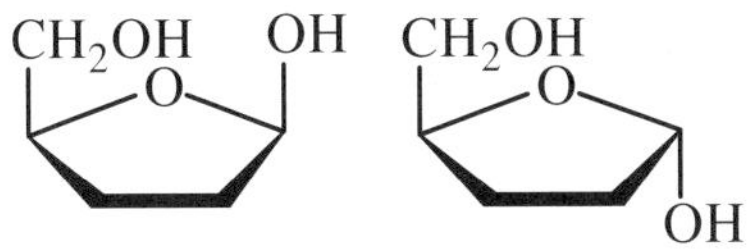

17.

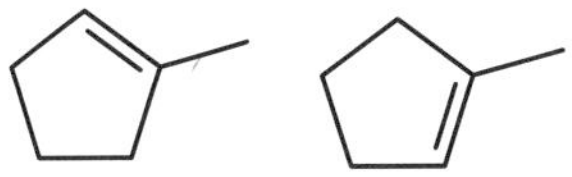

Short Answer

18. Write the equation for determining specific rotation and define all variables that appear.

19. An equal mixture of enantiomers is called _____.

20. Draw a Fischer projection of the molecule below. Place the carboxylic acid group at the top and the carbons in an apparent vertical line.

21. What is the maximum number of stereoisomers one should expect for a structure having four stereogenic centers?

22. What is meant by the term dextrorotatory as it applies to polarimetry?

23. A compound containing two stereogenic centers has a melting point of +39°C. What can you predict about the melting point of its enantiomer?

24. Consider a diastereomer of the compound mentioned in problem 23 above. What can you predict about its melting point?

25–28. Assign the absolute configuration R or S to each molecule below.

Chapter Test: Answers

1. **F** 2. **T** 3. **T** 4. **F** 5. **a** 6. **e** 7. **b** 8. **c** 9. **d** 10. **a** 11. **b** 12. **a**
13. **c** 14. **d** 15. **c** 16. **d** 17. **a**

18. $[\alpha]_D = \dfrac{\alpha}{(l \times C)}$

where $[\alpha]_D$ is the specific rotation at 589 nm (commonly called the sodium D line); α is the measured rotation, which is an experimentally generated value expressed in degrees; l is the path length through which light passes measured in decimeters; and C is the concentration expressed in g/mL.

19. a racemic mixture

20. [Fischer projection with CO₂H at top, HO—|—H, then C≡C, then H at bottom]

21. 16

22. Compounds that rotate plane-polarized light to the right (clockwise).

108   Chapter 5   Stereochemistry

23. Enantiomers have identical melting points so it will be +39°C.

24. Although the melting points might be close, this is not necessarily true and the melting point of this diastereomer could be quite different. The melting point will likely not be +39°C unless it is purely coincidental.

25. *R*

26. *S*

27. *R*

28. *S*

Check Your Performance

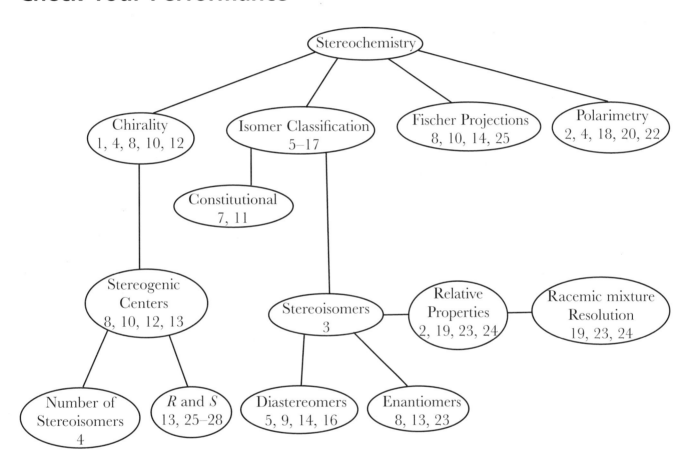

Midterm Exam

True/False

1. Only carbons that are double bonded are sp² hybridized.
2. Hexane and cyclohexane are isomers.
3. Ozonolysis of 2-butyne yields 2 equivalents of CH_3CO_2H
4. There are no chiral compounds with the formula C_4H_6.
5. Resolution of a racemic mixture is the measuring of its optical rotation in a polarimeter.

Multiple Choice

6. Which is a correct resonance form of $H-\overset{\ominus}{\underset{H}{C}}-\overset{\oplus}{N}\equiv N:$?

   a. $:N\equiv\overset{\oplus}{N}-\overset{\ominus}{\underset{H}{\ddot{C}}}-H$

   b. $H-\overset{\oplus}{\underset{H}{\ddot{C}}}-\overset{\ominus}{N}\equiv N:$

   c. $H-C\equiv\overset{\oplus}{N}-\overset{\ominus}{\ddot{N}}-H$

   d. $H-\underset{H}{C}=\overset{\oplus}{N}=\overset{\ominus}{\ddot{N}}:$

   e. None of the above

7. Which of the following has the highest boiling point?
   a. Hexane
   b. 2-Methylpentane
   c. 2,2-Dimethylbutane
   d. 2,3-Dimethylbutane
   e. All of the above are isomers and therefore have the same boiling point.

8. The substituents on the most stable conformer of a *cis*-1,3 disubstituted cyclohexane will be
   a. both axial.
   b. both equatorial.
   c. one axial and one equatorial.
   d. on opposite faces of the ring.
   e. None of the above

9. Which of the following can exist as *cis* or *trans*?
   a. Methylcyclobutane
   b. 1,1-Dimethylcyclobutane
   c. 1,2-Dimethylcyclobutane
   d. 1-Butene
   e. All of the above

10. Which group below has the highest CIP priority?
    a. —C≡N
    b. —CH$_2$NH$_2$
    c. —C(CH$_3$)$_3$
    d. —CH$_2$CH$_2$CH$_2$CH$_3$
    e. —CH=NH

Select one of the following terms to describe the relationships for each of the pairs of structural formulas shown in 11 through 15.
   a. Enantiomers
   b. Diastereomers
   c. Constitutional isomers
   d. Same
   e. None of the above

11. CH$_3$C(OH)=CH$_2$ CH$_3$CCH$_3$ (with C=O)

12.

13.

14.

15.

Short Answer

For each of problems 16 through 22 provide an unambiguous structural formula for the organic product obtained from treating 1-ethylcyclobutene with the indicated reagents.

Midterm Exam   111

16. Br$_2$, CCl$_4$

17. Br$_2$, H$_2$O

18. Hg(O$_2$CCH$_3$)$_2$, H$_2$O, THF, then NaBH$_4$

19. BH$_3$, then H$_2$O$_2$ and hydroxide

20. OsO$_4$, then aqueous NaHSO$_3$

21. Hot acidic aqueous potassium permanganate

22. H$_2$, Pd

For each of problems 23 through 25 provide an unambiguous structural formula for the organic product obtained from treating 3-methyl-1-butyne with the indicated reagents.

23. Excess HBr

24. BH$_3$, then alkaline H$_2$O$_2$

25. NaNH$_2$, then CH$_3$CH$_2$CH$_2$Br

Midterm Exam Answers

1. **F** 2. **F** 3. **T** 4. **T** 5. **F** 6. **d** 7. **a** 8. **b** 9. **c** 10. **a** 11. **c** 12. **b**
13. **b** 14. **d** 15. **a**

16. 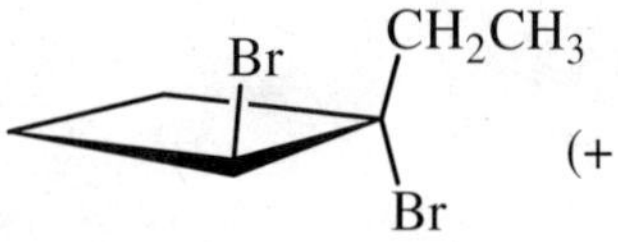 [structure: cyclohexane with Br, Br, CH$_2$CH$_3$] (+ enantiomer)

17. 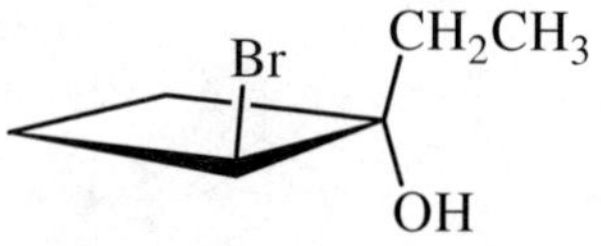 [structure: cyclohexane with Br, OH, CH$_2$CH$_3$] (+ enantiomer)

18. 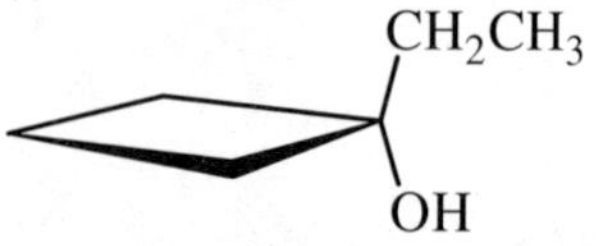 [structure: cyclohexane with CH$_2$CH$_3$, OH]

19. 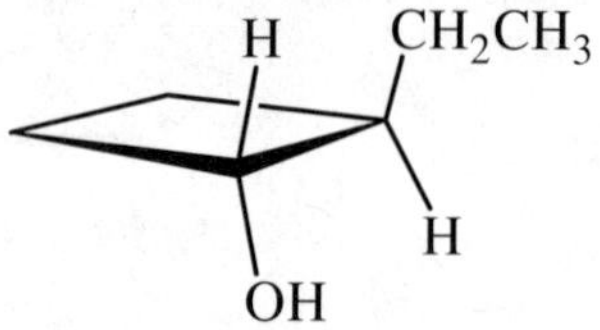 [structure: cyclohexane with H, OH, CH$_2$CH$_3$, H] (+ enantiomer)

20. 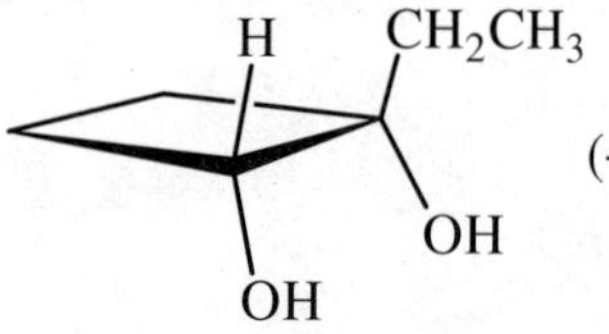 [structure: cyclohexane with H, OH, OH, CH$_2$CH$_3$] (+ enantiomer)

21. 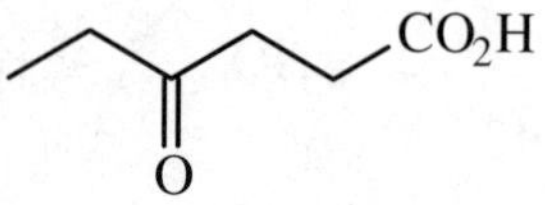 [structure: CH$_3$CH$_2$-C(=O)-CH$_2$CH$_2$-CO$_2$H]

22. 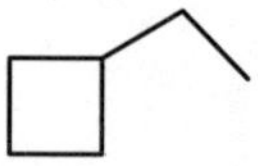 [structure: cyclobutane with methyl group]

23. (CH$_3$)$_2$CHCBr$_2$CH$_3$

24. (CH$_3$)$_2$CHCH$_2$CHO

25. (CH$_3$)$_2$CHC≡CCH$_2$CH$_2$CH$_3$

CHAPTER 6

Alkyl Halides: Substitution and Elimination Mechanisms

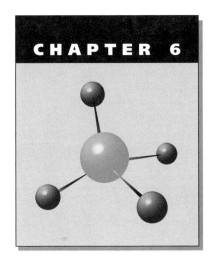

Naturally occurring compounds containing C—X bonds are mostly found in marine organisms. Many synthetic organic halogen compounds are important in industry, with over 15,000 produced commercially. Organic chemists use alkyl halides to prepare a variety of other kinds of compounds. In this chapter we survey some transformations these versatile compounds can undergo.

ESSENTIAL BACKGROUND

- Electronegativity and bond polarity (Chapter 1)
- Resonance (Chapter 1)
- Nomenclature of alkanes and alkyl groups (Chapter 2)
- Conformations and dihedral angle (Chapter 2)
- Radical halogenation of alkanes (Chapter 2)
- Writing reaction mechanisms (Chapters 2 and 3)
- Identifying nucleophiles and electrophiles (Chapter 3)
- Addition of HX or X_2 to alkenes (Chapter 3)
- Zaitsev's rule for elimination yielding alkenes (Chapter 3)

TOPIC 1: STRUCTURES, NOMENCLATURE, AND PROPERTIES OF ALKYL HALIDES

KEY POINTS

✓ *How are alkyl halides named?*

✓ *What is the polarity of the C—X bond?*

✓ *How do the melting point, boiling point, and density of RX compare to those of alkanes?*

Alkyl halides have the general structure RX, where R is alkyl group and X is a halogen. The International Union of Pure and Applied Chemistry (IUPAC) names for alkyl halides are derived as halo-substituted parents. The common (trivial) names are two words with the first being the name of the alkyl group and the second the name of the halide. **Table 6.1** shows several examples with both the systematic IUPAC names and the common names.

Table 6.1 Structures and Names for Some Alkyl Halides

STRUCTURAL FORMULA	IUPAC NAME	COMMON NAME
CH_3I	Iodomethane	Methyl iodide
CH_3CHFCH_3	2-Fluoropropane	Isopropyl fluoride
$CHCl_3$	Trichloromethane	Chloroform
CH_2Cl_2	Dichloromethane	Methylene chloride
$BrCH_2CH_3$	Bromoethane	Ethyl bromide
$ICH_2CH(CH_3)_2$	1-Iodo-2-methylpropane	Isobutyl bromide
$(CH_3)_3CBr$	2-Bromo-2-methylpropane	*tert*-Butyl bromide

The C—X bonds of alkyl halides are polarized toward the halogen. Alkyl halides are generally more polar than alkanes of comparable size, and they have higher boiling points, melting points, and densities.

Topic Test 1: Structures, Nomenclature, and Properties of Alkyl Halides

True/False

1. The common or trivial names of alkyl halides are usually one word.
2. The boiling point of propane is higher than that of 1-chloropropane.

Multiple Choice

3. The systematic IUPAC name for chloroform is
   a. carbon tetrachloride.
   b. methyl chloride.
   c. chloromethane.
   d. trichloromethane.
   e. None of the above

4. The systematic IUPAC name for *tert*-butyl bromide is
   a. tribromobutane.
   b. 1-bromobutane.
   c. 1-bromo-1,1-dimethylethane.
   d. 2-bromo-2-methylpropane.
   e. None of the above

Short Answer

5. How many iodine atoms are contained in a molecule of methylene iodide?
6. Name the following according to IUPAC conventions.

[Structure: cyclohexane with F and CH₂CH₂CH₃ on same carbon]

Topic Test 1: Answers

1. **False.** The common or trivial names of alkyl halides are usually *two* words. For most simple alkyl monohalides, the first word is the name of the alkyl group and the second word is the name of the halogen in its anionic form (fluoride, chloride, bromide, or iodide).

2. **False.** Generally, alkyl halides have higher boiling points than their alkane parents due to the increase in size and polarity that accompanies replacement of a hydrogen with a halogen.

3. **d**

4. **d.** The longest parent chain is three carbons, and a bromo and a methyl substituent (listed alphabetically) are found on the second carbon of the propane chain.

5. **Two.** The methylene group is a CH_2 that requires two halogen atoms to give four bonds around carbon.

6. 1-Fluoro-1-propylcyclohexane. Note the spelling of the halogen ("u" before "o"). The substituents are listed alphabetically, and the numbers are required to specify that the substituents are both on the same ring carbon.

TOPIC 2: PREPARATION OF ALKYL HALIDES

KEY POINTS

✓ *How are alkyl monohalides and vicinal dihalides made from alkenes?*

✓ *How are alkyl halides prepared from alkanes?*

✓ *How are alkyl halides prepared from alcohols?*

We saw in Chapter 3 that alkenes react with HX or X_2 to yield alkyl halides and vicinal dihalides respectively.

[Reaction scheme: C=C + HX → C(H)-C(X); C=C + X₂ → C(X)-C(X)]

Recall from Chapter 2 that radical substitution reactions of alkanes with X_2 are sometimes used to make alkyl halides. Brominations are more selective than chlorinations, with the former yielding predominantly the most highly substituted alkyl bromide and the former giving complex mixtures where possible.

$$CH_3CH_3 \xrightarrow[h\nu]{X_2\ (X = Cl,\ Br)} CH_3CH_2X$$

$$\underset{\underset{CH_3}{|}}{CH_3CHCH_3} \xrightarrow{Br_2} \underset{\underset{CH_3}{|}}{\overset{\overset{Br}{|}}{CH_3CCH_3}}$$

Selective free radical brominations can be performed with *N*-bromosuccinimide (NBS). The reaction intermediate is a resonance-stabilized **allylic radical** that can sometimes yield more than one **allylic bromination** product.

Alkyl halides are often prepared from alcohols by treatment with HX, PBr$_3$, or SOCl$_2$. Reactions with HX work best for tertiary alcohols where less substituted alcohols are more efficiently converted using the latter reagents.

$$\underset{\underset{R}{|}}{\overset{\overset{R}{|}}{R-C-OH}} \xrightarrow{HX} \underset{\underset{R}{|}}{\overset{\overset{R}{|}}{R-C-X}}$$

$$RCH_2Br \xleftarrow{PBr_3} RCH_2OH \xrightarrow{SOCl_2} RCH_2Cl$$

$$\underset{\underset{Br}{|}}{RCHR} \xleftarrow{PBr_3} \underset{\underset{OH}{|}}{RCHR} \xrightarrow{SOCl_2} \underset{\underset{Cl}{|}}{RCHR}$$

Topic Test 2: Preparation of Alkyl Halides

True/False

1. NBS can be used to selectively prepare allylic alkyl bromides.
2. Ethanol, CH$_3$CH$_2$OH can be converted to bromoethane with PBr$_3$.

Multiple Choice

3. Which of the following is a method for synthesizing alkyl halides?
   a. HX + alkane
   b. HX + alkene
   c. Alcohol + X_2
   d. All of the above
   e. None of the above

4. What is the best set of conditions for the following conversion? $CH_3OH \rightarrow CH_3Cl$
   a. HCl
   b. $SOCl_2$
   c. Cl_2 and hv
   d. PBr_3
   e. NBS

Short Answer

5. Show how one could convert 1-ethylcyclopentene into 1-bromo-1-ethylcyclopentane.

6. Show the reagents and/or conditions one could use to carry out the transformation below.

Topic Test 2: Answers

1. **True.** Under free radical conditions, the mechanism proceeds via a resonance-stabilized allylic radical.

2. **True.** Phosphorus tribromide, PBr_3, is the reagent of choice for converting a primary alcohol to an alkyl bromide. Hydrogen bromide works best when the starting alcohol is tertiary.

3. **b**

4. **b.** Although a appears reasonable, that reagent works best if the alcohol is tertiary. Reagents c are for alkane chlorination. Choices d and e are both absurd in that besides not reacting with alcohols in any way we have seen, they are also brominating agents and are used for alkyl bromide synthesis.

5. **HBr.** The product results from HBr addition across the alkene to give Markovnikov orientation.

6. **Br_2.** Recall that the bromine atoms add across the alkene from opposite sides, yielding a *trans* dibromide.

TOPIC 3: NUCLEOPHILIC SUBSTITUTION REACTIONS: S_N1 AND S_N2

KEY POINTS

✓ *What is the general form of a nucleophilic substitution reaction?*

✓ What do the terms "substrate" and "leaving group" mean in these reactions?

✓ What are the mechanisms for nucleophilic substitution?

✓ What factors control which substitution mechanism will operate?

✓ What are the characteristics and consequences of S_N2 and S_N1 mechanisms?

Alkyl halides react with nucleophiles to yield substitution products in which the nucleophile replaces the halogen. In these reactions, RX is called the **substrate** and the halide ion is referred to as a **leaving group**. There are two major mechanistic pathways for these reactions: S_N1 and S_N2.

$$Nu^- + R{-}X \rightarrow Nu{-}R + X^-$$

The S_N2 process (substitution, nucleophilic, bimolecular) is a concerted one-step reaction that has no intermediates, passes through one transition structure, and leads to **inversion** of configuration around the substrate's reactive site. Kinetic experiments show that the reaction rate depends on the concentration of both the nucleophile and the substrate:

$$\text{Rate} = k[Nu][RX]$$

The attack of the nucleophile is from the "back side" and occurs simultaneous to the breaking of the C—X bond. Steric hindrance by alkyl groups prevents back side attack so S_N2 reactions are favored for substrates in the order $CH_3 > 1° > 2° \gg 3°$. Polar aprotic solvents such as dimethylsulfoxide, dimethylformamide, or acetonitrile favor these reactions.

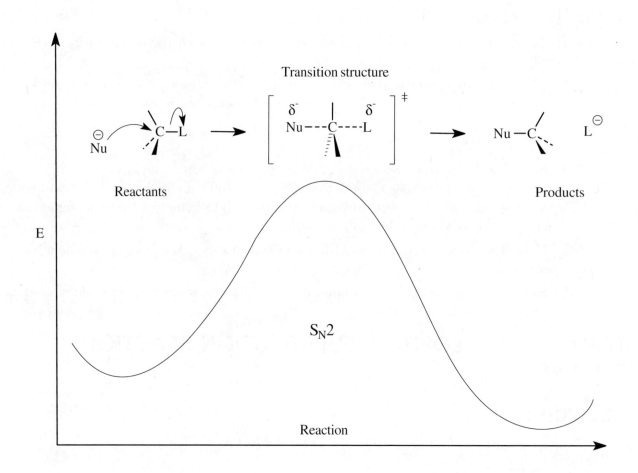

The S_N1 mechanism (substitution, nucleophilic, unimolecular) is a two-step process that occurs via a carbocation intermediate and passes through two transition structures. The first step is **rate limiting** (i.e., the slowest) and is a **solvolysis** where solvent helps the substrate to ionize into a carbocation and leaving group. In the second step, the nucleophile can attack either side of the planar carbocation intermediate, which generally leads to products where the configuration around the reactive site is scrambled to a mixture of stereoisomers if possible. The reaction rate depends only on the concentration of the substrate and is independent of nucleophile concentration: Rate = k[RX]. Because there is a carbocation intermediate, the S_N1 mechanism is favored according to the stability of carbocations, $CH_3 \ll 1° < 2° < 3°$, and is favored by polar solvents that stabilize carbocations.

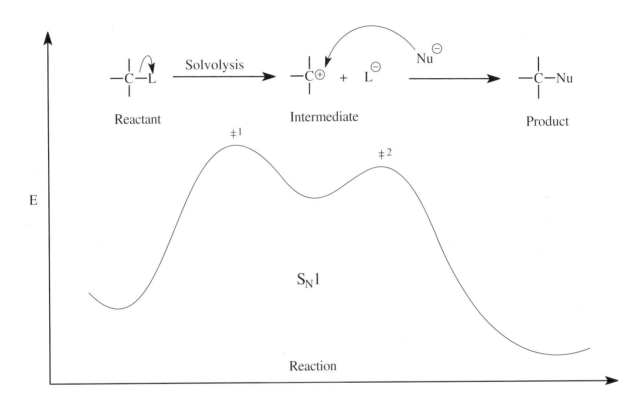

Topic Test 3: Nucleophilic Substitution Reactions: S_N1 and S_N2

True/False

1. An S_N2 reaction has two mechanistic steps.
2. An S_N1 reaction will likely be faster for a 3° than for a 1° R—X.

Multiple Choice

3. In a nucleophilic substitution on an alkyl halide, the halogen will usually be the
   a. nucleophile.
   b. substrate.

    c. leaving group.
    d. solvent.
    e. None of the above.

 4. If (R)-2-bromobutane were heated with sodium iodide in polar solvent resulting in an S_N1 reaction, the product would be
    a. (S)-2-bromobutane.
    b. (R)-2-iodobutane.
    c. (S)-2-iodobutane.
    d. racemic 2-iodobutane.
    e. None of the above

Short Answer

 5. Show how one could use nucleophilic substitution to make $CH_3CH_2OCH_2CH_3$ from bromoethane and a suitable nucleophile.

 6. By what mechanism do you predict your reaction in number 5 above occurs? Explain.

Topic Test 3: Answers

1. **False**

2. **True**

3. **c.** The general reaction is $Nu^- + R-X \rightarrow Nu-R + X^-$, which shows the halide "leaving group" being replaced by the nucleophile.

4. **d.** The S_N1 mechanism involves a planar carbocation intermediate that has a 50/50 likelihood of being attacked by the iodide nucleophile.

5. $CH_3CH_2O^- + BrCH_2CH_3 \rightarrow CH_3CH_2OCH_2CH_3 + Br^-$

6. The reaction in problem 5 above will likely occur by the S_N2 mechanism because the alkyl halide is primary and so back side attack is not sterically hindered. Also, the primary carbocation required for an S_N1 reaction would be unstable and difficult to form.

TOPIC 4: ELIMINATION REACTIONS: E1 AND E2

KEY POINTS

 ✓ *What are the mechanisms for elimination?*

 ✓ *What factors control which elimination mechanism will operate?*

 ✓ *What are the characteristics and consequences of E2 and E1 mechanisms?*

Recall from Chapter 3 that dehydrohalogenation yields alkenes from alkyl halides.

$$-\underset{H}{\overset{|}{C}}-\underset{X}{\overset{|}{C}}- \xrightarrow[(-HX)]{\text{Base}} -\overset{|}{C}=\overset{|}{C}-$$

There are two possible pathways whereby these reactions occur: E2 and E1. The E2 mechanism (elimination, bimolecular) is a concerted one-step process. There are no intermediates and one transition structure. The base abstracts the hydrogen at the same time the leaving group detaches. The preferred geometry of the transition structure is **anti-periplanar** (dihedral angle = 180°), although **syn-periplanar** is possible (dihedral angle = 0°).

The rate of reaction depends on the concentrations of both base and substrate:

$$\text{Rate} = k[\text{RX}][\text{B}^-]$$

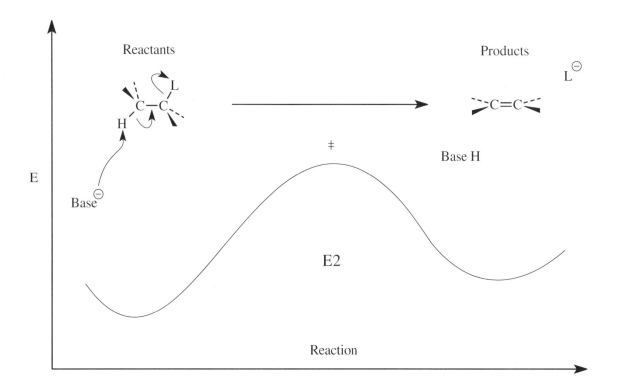

The E1 mechanism (elimination, unimolecular) is a two-step process involving a carbocation intermediate and two transition structures. Solvolysis removes the leaving group in the rate-limiting first step and then the base removes a proton from a carbon adjacent to the cationic center. The rate depends only on concentration of the substrate and is independent of the base concentration. Rate = k[RX]. Because there is a carbocation intermediate, the E1 mechanism is favored according to the stability of carbocations, $CH_3 \ll 1° < 2° < 3°$, and is favored by polar solvents that stabilize carbocations.

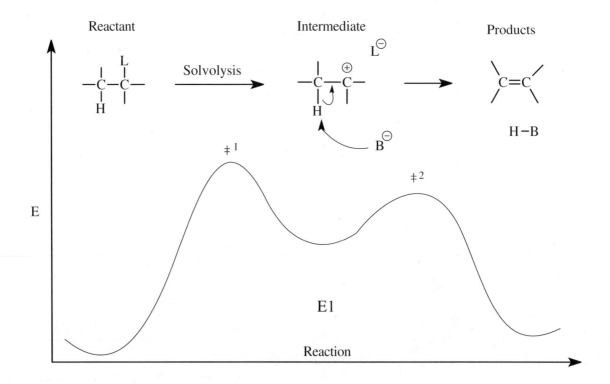

Topic Test 4: Elimination Reactions: E1 and E2

True/False

1. The rate of an E2 reaction will be independent of the base concentration.
2. The two "humps" in the energy versus reaction diagram for an E1 reaction correspond to the base and the leaving group.

Multiple Choice

3. The first step in the E1 mechanism
   a. is solvolysis.
   b. is the same as that of the S_N1 mechanism.
   c. yields a carbocation.
   d. All of the above
   e. None of the above

4. An E2 reaction of an alkyl halide
   a. occurs in two steps.
   b. occurs via a carbocation.
   c. will yield an alkene.
   d. All of the above
   e. None of the above

Short Answer

5. Assume that (S)-2-bromobutane is treated with a strong base, resulting in an E2 reaction. Draw the conformations of starting material that lead to *cis*- and *trans*-2-butene via the *anti*-periplanar transition structures.

6. Assume (*S*)-2-iodobutane is treated with a weak base in polar solvent, resulting in an E1 reaction to give 2-butene. Describe the stereochemical outcome of the reaction and explain your reasoning.

Topic Test 4: Answers

1. **False.** Because the base is involved in the rate-limiting (and only) step in the mechanism, the base concentration does effect the rate of reaction.

2. **False.** Although the energy versus reaction curve does show two maxima, they correspond to transition structures (often called transition states) leading to and from the carbocation intermediate.

3. **d.** The solvolysis that leads to a carbocation in the first step of E1 is the same as the process that occurs as the first step of the S_N1 mechanism.

4. **c.** This is dehydrohalogenation (Chapter 3).

5.

6. The solvolysis will yield a planar carbocation with free rotation around the C(2)—C(3) bond. When the base abstracts a proton from adjacent to the cation center, the relative positions of the methyl groups at that time will be "locked in" and manifested as *cis*- or *trans*-2-butene. Because the conformer of carbocation that leads to the *trans* product is lower energy than the conformer leading to *cis*, one expects more *trans* than *cis* product.

TOPIC 5: COMPETING REACTIONS: SUMMARY OF S_N1, S_N2, E1, AND E2

Key Points

✓ Can the product(s) of a substitution/elimination reaction of RX be predicted?

✓ By what mechanism will a given reaction occur?

A given nucleophile could also act as a base and vice versa. Predicting whether a given pair of reagents will react to give substitution and/or elimination or by which of the four mechanisms the reaction will proceed is not always easy. There are, however, some cases that are easier than others to generalize based on carbocation stability, base strength, and so on. **Methyl halides** undergo S_N2 reactions but not S_N1, E2, or E1. **Primary** alkyl halides react by S_N2 or E2 but not S_N1 or E1. The E2 process is favored over S_N2 if a strong and/or bulky base is present such as *tert*-butoxide. **Secondary** alkyl halides can react by any of the four mechanisms. Strong and/or bulky bases favor E2 reactions. S_N2 reactions often occur in competition with E2. Unimolecular reactions are favored for secondary carbocations that are resonance stabilized such as allylic or benzylic systems. **Tertiary** alkyl halides do not undergo S_N2 reactions but can react by S_N1, especially in protic solvents. Elimination can occur by E2 or E1 with the latter usually in competition with S_N1.

Topic Test 5: Competing Reactions: Summary of S_N1, S_N2, E1, and E2

True/False

1. Elimination reactions often compete with nucleophilic substitution reactions.

2. Secondary alkyl halides do not undergo S_N1 or E2 reactions.

Multiple Choice

3. What is the most likely mechanism for this reaction?

   a. S_N2
   b. S_N1
   c. E2
   d. E1
   e. None of the above

4. What is the most likely mechanism for this reaction?

$$(CH_3CH_2)_3N + CH_3CH_2I \rightarrow [(CH_3CH_2)_4N^+]I^-$$

    a. S$_N$2
    b. S$_N$1
    c. E2
    d. E1
    e. None of the above

Short Answer

Provide unambiguous structural formulas for the missing organic products.

5. CH$_3$I + NaOH →

6.
$$(CH_3)_3C-Br \xrightarrow[HOCH_2CH_3]{Na^{\oplus} \ ^{\ominus}OCH_2CH_3}$$

Topic Test 5: Answers

1. **True.** Many nucleophiles can also act as bases and vice versa.
2. **False.** Secondary alkyl halides can react by any of the mechanisms from this chapter.
3. **c.** The alkene product indicates elimination. Without more information it is difficult to rigorously rule out E1 but the strong bulky base *t*-butoxide suggests E2.
4. **a.** The structure of the product indicates substitution rather than elimination. The primary alkyl halide is not sterically hindered and would give an unstable carbocation and therefore will substitute by S$_N$2 and not by S$_N$1.
5. **CH$_3$OH.** Of the four mechanisms discussed in this chapter, only S$_N$2 is possible for this methyl halide starting material.
6. **(CH$_3$)$_2$C=CH$_2$.** The tertiary substrate, strong base, and polar solvent suggest the major mechanism is E2. S$_N$2 is not possible due to steric hindrance to back side attack. Some product resulting from S$_N$1 may be observed as a minor product, i.e., (CH$_3$)$_3$COCH$_2$CH$_3$.

TOPIC 6: ORGANOMETALLIC REAGENTS FROM ALKYL HALIDES

KEY POINTS

✓ *What are Grignard reagents and how are they prepared?*

✓ *What products result from reactions of Grignard reagents with acid?*

✓ *What are alkyllithium reagents and how are they prepared?*

✓ *What are Gilman reagents and how are they prepared?*

✓ *What products result from Gilman reagents reacting with alkyl halides?*

Alkyl halides react with magnesium metal in ether solvent to form alkylmagnesium halides that are commonly called Grignard reagents. The bond between the carbon and magnesium is strongly polarized in the direction of carbon such that a Grignard reagent acts as if it is a carbanion (negatively charged carbon species). Grignard reagents are good nucleophiles and are also fairly basic. They react with electrophiles and acids accordingly.

$$RX \xrightarrow[\text{Ether}]{\text{Mg}} RMgX \quad = \quad \overset{\delta^-}{R}\text{—}\overset{\delta^+}{MgX} \quad (\text{Acts like } R^{\ominus})$$

$$RMgX \xrightarrow{H_2O} RH \quad (+ \; MgXOH)$$

Alkyl halides react with lithium metal in cold pentane or ether to yield alkyllithium reagents RLi. Like Grignard reagents, alkyllithiums are organometallic and have a reactivity one would expect for a carbanion.

$$RX \xrightarrow[\text{Pentane}]{\text{2 Li}} RLi \quad + \quad LiX$$

$$RLi \quad = \quad \overset{\delta^-}{R}\text{—}\overset{\delta^+}{Li} \quad (\text{Acts like } R^{\ominus})$$

Alkyllithiums are sometimes difficult to handle due to their high reactivity. When they are treated with CuI in ether, the products formed are lithium diorganocopper compounds called Gilman reagents. These have a variety of uses, including their reaction with alkyl halides to give coupling products.

$$2 \; RLi \quad + \quad CuI \xrightarrow{\text{Ether}} (R_2Cu)^{\ominus} Li^{\oplus} \quad + \quad LiI$$
$$\text{Gilman reagent}$$

$$(R_2Cu)^{\ominus} Li^{\oplus} \quad + \quad R'X \xrightarrow{\text{Ether}} R\text{—}R' \quad (+ \; LiX + RCu)$$

All the organometallic reactions above apply to other organohalogen compounds, including alkenyl and aryl halides in addition to simple alkyl halides.

$$CH_2\text{=}CH\text{—}Br \xrightarrow[\text{Ether}]{\text{Mg}} CH_2\text{=}CH\text{—}MgBr \quad \text{Vinyl Grignard}$$

$$\text{Ph—I} \xrightarrow[\text{Pentane}]{\text{Li}} \text{Ph—Li} \xrightarrow{\text{CuI}} [(\text{Ph})_2 Cu]^{\ominus} Li^{\oplus} \xrightarrow{CH_2\text{=}CHBr} \text{Ph—CH=CH}_2$$

Topic Test 6: Organometallic Reagents from Alkyl Halides

True/False

1. Alkyllithium compounds behave as though they were carbocations.
2. Grignard reagents react like bases in the presence of a proton source.

Multiple Choice

3. Which is true about Gilman reagents?
   a. They are prepared from Grignard reagents.
   b. They are prepared from alkyllithiums.
   c. They are used to prepare alkyl halides.
   d. All of the above
   e. None of the above

4. If CH_3CH_2Li were treated with water, the result would be
   a. $CH_3CH_2OH + LiH$.
   b. $CH_3CH_3 + LiOH$.
   c. $CH_3CH_2CH_2CH_3 + Li$.
   d. No reaction
   e. None of the above

Short Answer

5. Show how one could prepare butane using reactions from this topic and using bromoethane as the only carbon source. More than one step may be required.

6. Show how one could convert 1-bromopropane to propane in two steps via a Grignard reagent.

Topic Test 6: Answers

1. **False.** Alkyllithiums behave as though they were carbanions.
2. **True.** These carbanion-like reagents are essentially the conjugate bases of alkanes. They react completely with water or other proton donors to yield the protonated form.
3. **b**

$$2\,RLi + CuI \xrightarrow{ether} \underset{\text{Gilman reagent}}{(R_2Cu^-)Li^+} + LiI$$

4. **b.** An acid-base reaction of the carbanion like RLi.

5. $CH_3CH_2Br \xrightarrow{Li} CH_3CH_2Li \xrightarrow{CuI} [(CH_3CH_2)_2Cu]Li \xrightarrow{CH_3CH_2Br} CH_3CH_2CH_2CH_3$

6. $CH_3CH_2CH_2Br \xrightarrow{Mg/ether} CH_3CH_2CH_2MgBr \xrightarrow{H_2O} CH_3CH_2CH_2CH_3$

APPLICATION

Several polychlorinated organic compounds have been shown to be effective insecticides. Although their toxicity to humans is generally quite low, some of these have the undesirable property of persisting in the environment where they can accumulate and adversely affect birds and fish. DDT, for example, has been banned in the United States since 1972.

DDT   Lindane   Chlordane

DEMONSTRATION PROBLEM

Show how one could synthesize butylcyclopentane using alcohols with five or fewer carbons as the only carbon source.

Solution

Reasoning backward from the target compound suggests that this C_9 alkane was probably built from $C_5 + C_4$ fragments. The best choice for the "new" bond that was formed is the point where the alkyl group is attached to the ring. One way to form a C—C bond is with a Gilman reagent and an alkyl halide. There are two reasonable routes to the product.

$[(C_5H_9)_2Cu]^- Li^+$ + $XCH_2CH_2CH_2CH_3$

or

$[(CH_3CH_2CH_2CH_2)_2Cu]^- Li^+$ + cyclopentyl–X

The second path is arbitrarily selected here. The remainder of the problem is to make the RX and R_2CuLi from alcohols. There are several possible correct methods for converting ROH into RX, but recall, however, that HX works well only for tertiary alcohols and our alcohols are primary and secondary.

128 Chapter 6 Alkyl Halides: Substitution and Elimination Mechanisms

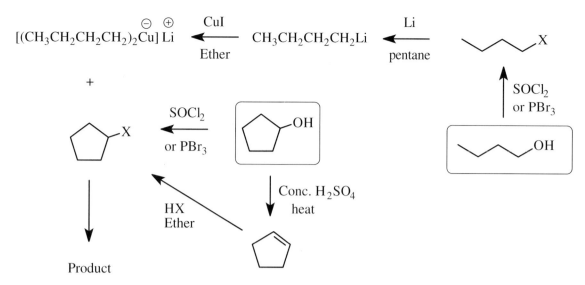

Chapter Test

True/False

1. Iodomethane is more dense than ethane.
2. Grignard reagents (RMgX) react with acids to give alkanes (RH).
3. Tertiary alkyl halides do not react by the E2 mechanism.
4. The general formula for a Gilman reagent is RCuX.

Multiple Choice

5. Carbon number one of 1-bromopropane
    a. bears a partial negative charge.
    b. bears a partial positive charge.
    c. acts as a nucleophile.
    d. acts as a leaving group.
    e. None of the above

6. Which reagent below is often used to convert an alkene into an alkyl halide?
    a. Thionyl chloride, $SOCl_2$
    b. Phosphorus tribromide, PBr_3
    c. Cuprous iodide (CuI) in ether
    d. HBr
    e. None of the above

7. The reagent NBS is used for
    a. preparing Gilman reagents.
    b. preparing Grignard reagents.
    c. preparing alkenes out of alkyl halides.
    d. bromination of allylic positions.
    e. None of the above

8. When the concentration of nucleophile is doubled, the rate of an S_N1 reaction
    a. doubles.
    b. stays the same.

c. slows to half.
d. falls to zero.
e. None of the above

9. Which compound could be most cleanly made by free-radical chlorination of an alkane?
   a. 1-Chloropentane
   b. 1-Chloro-3-methylbutane
   c. 1-Chloro-2-methylbutane
   d. 2-Chloro-3-methylbutane
   e. 1-Chloro-2,2-dimethylpropane

Short Answer

Name the following compounds. Where possible include both the common and systematic IUPAC names.

10. $H_2C=CHCH_2Br$

11.

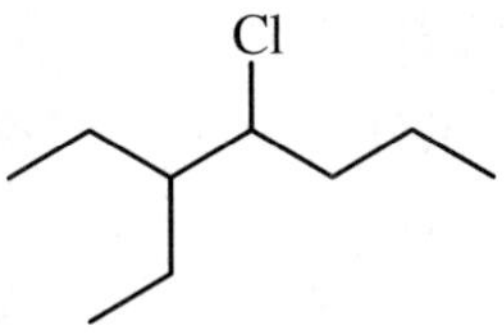

12–13. Show how one could prepare each of the following compounds from an alkene.

12.

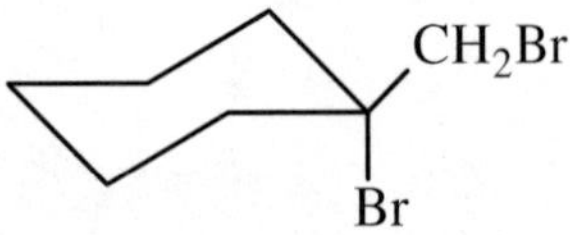

13.

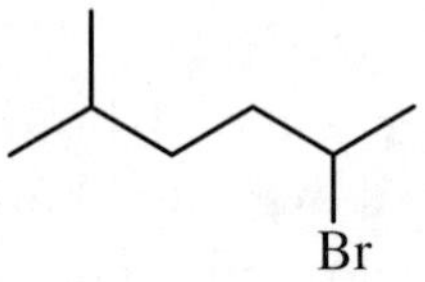

14–16. Show how one could prepare each of the following from an alcohol.

14. $CH_3CH_2\underset{\underset{CH_3}{|}}{\overset{\overset{CH_3}{|}}{C}}-Cl$

15.

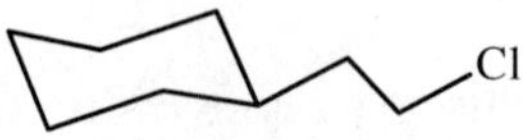

16.

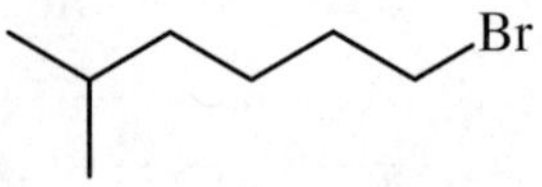

17–19. Show how to prepare each of the following from cyclohexene.

17.

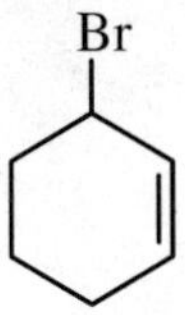

18.

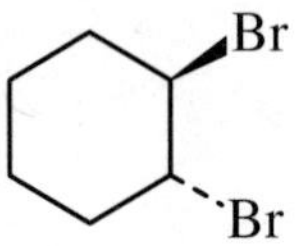

19.

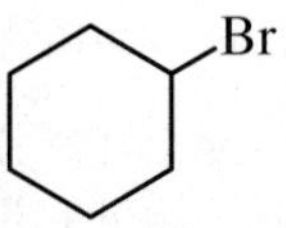

For each item below, specify S_N1, S_N2, E1, or E2.

20. Back side attack

21. Periplanar transition structure

22. Inversion of configuration

23. Racemization of configuration

24. Yields alkene in a single step
25. Most likely when the substrate is a methyl iodide
26. Produces an alkene via a carbocation
27. Rate of this alkene-producing reaction is independent of base concentration.
28. $(CH_3)_3CBr$ is heated in $HOCH_2CH_3$ to yield $(CH_3)_3COCH_2CH_3$.
29. Show how to prepare the compound below from bromocyclohexane.

30. Assume D_2O is available as a deuterium source. Show how one could prepare CH_3CH_2D from bromoethane via a Grignard reagent.

Chapter Test Answers

1. **T** 2. **T** 3. **F** 4. **F** 5. **b** 6. **d** 7. **d** 8. **b** 9. **e**
10. Allyl bromide or 3-bromopropene
11. 4-Chloro-3-ethylheptane

12. [methylenecyclohexane + Br₂ → 1-bromo-1-(bromomethyl)cyclohexane]

13. [4-methyl-1-pentene + HBr → 2-bromo-4-methylpentane]

14. $CH_3CH_2\underset{\underset{CH_3}{|}}{\overset{\overset{CH_3}{|}}{C}}-OH$ $\xrightarrow{HCl}$ $CH_3CH_2\underset{\underset{CH_3}{|}}{\overset{\overset{CH_3}{|}}{C}}-Cl$

15. [cyclohexylmethanol (CH₂OH) + SOCl₂ → cyclohexylmethyl chloride (CH₂Cl)]

16. [4-methyl-1-pentanol + PBr₃ → 1-bromo-4-methylpentane]

17. [cyclohexene + NBS → 3-bromocyclohexene]

18. [cyclohexene + Br₂ → trans-1,2-dibromocyclohexane]

19. [cyclohexene] —HBr→ [cyclohexyl bromide]

20. S_N2
21. E2
22. S_N2
23. S_N1
24. E2
25. S_N2
26. E1
27. E1
28. S_N1

29.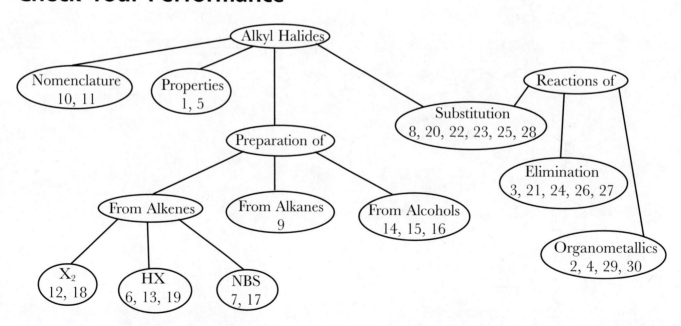

30. $CH_3CH_2Br \xrightarrow{Mg} CH_3CH_2MgBr \xrightarrow{D_2O} CH_3CH_2D$

Check Your Performance

Alkyl Halides
- Nomenclature 10, 11
- Properties 1, 5
- Preparation of
  - From Alkenes
    - X_2 12, 18
    - HX 6, 13, 19
    - NBS 7, 17
  - From Alkanes 9
  - From Alcohols 14, 15, 16
- Reactions of
  - Substitution 8, 20, 22, 23, 25, 28
  - Elimination 3, 21, 24, 26, 27
  - Organometallics 2, 4, 29, 30

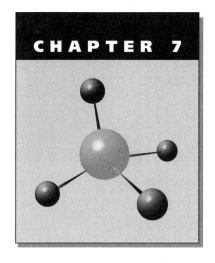

Dienes and Conjugation

CHAPTER 7

As the name implies, a diene contains two alkene linkages. The relationship between these two alkene groups can have a significant impact on the chemical nature of the diene. When the alkene groups are conjugated, the stability and reactivity of the diene is different from what is observed for nonconjugated dienes. How does one recognize conjugated and nonconjugated dienes? We answer that question in this chapter. First we examine the concept of conjugation in a general sense and then look specifically at dienes.

ESSENTIAL BACKGROUND

- Orbitals and hybridization (Chapter 1)
- Sigma and pi bonds (Chapter 1)
- Resonance (Chapter 1)
- Alkene nomenclature (Chapter 2)
- Electrophilic addition to alkenes (Chapter 2)
- Reaction mechanisms (Chapters 2 and 3)

TOPIC 1: CONJUGATION AND CONJUGATED DIENES

KEY POINTS

✓ *What does the term conjugation mean?*

✓ *How can one reliably identify the conjugated part of a molecule?*

✓ *What is a conjugated diene?*

✓ *What is the relative stability of conjugated and nonconjugated dienes?*

✓ *Do conjugated dienes react differently than simple alkenes?*

The term **conjugation** expresses a relationship between functional groups that appear to contain pi bonds, nonbonded electrons, or vacant orbitals. Molecules that contain several atoms over which electrons are resonance delocalized are described as **conjugated**. It is sometimes possible to identify conjugated systems by the appearance of alternating double and single bonds. This is, however, an oversimplification and can sometimes lead to misidentification. Illustrated below are some examples of systems that are conjugated yet in some cases do not appear to contain alternating double and single bonds. A more general and reliable identifier for conjugation is an orbital aligned for parallel overlap on three or more contiguous atoms. Note that this includes not only molecules that have alternating double and single bonds but also all the other conjugated systems.

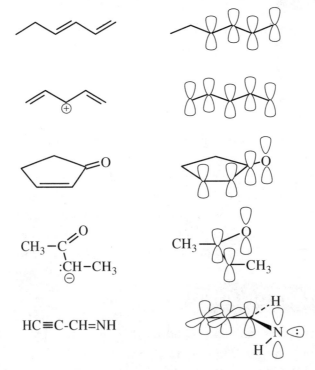

Conjugated dienes contain two adjacent alkene linkages and therefore have a parallel p orbital on each of four contiguous carbons. The simplest example is 1,3-butadiene. In general, conjugated dienes are more stable than those that are nonconjugated. Recall from Chapter 3 that most alkene chemistry takes place at the carbon-carbon pi bond. Like simple alkenes, conjugated dienes react at their pi bonds, but conjugated dienes undergo some reactions that do not apply to simple alkenes.

Topic Test 1: Conjugation and Conjugated Dienes

True/False

1. Isomerization of 1,4-cyclohexadiene to 1,3-cyclohexadiene is exergonic.
2. All conjugated molecules have alternating double and single bonds.

Multiple Choice

3. Which of the following is conjugated?
   a. $H_2C=C=CH_2$
   b. $H_2C=C=C=CH_2$
   c. $CH_3-C\equiv C-CH_3$
   d. None of the above
   e. All of the above

4. Which of the following ions is conjugated?
   a. $H_2C=C-CH_2^+$ (the allyl carbocation)
   b. $H_2C=C-CH_2CH_2^+$

c. :N≡C:⁻ (cyanide ion)
d. All of the above
e. None of the above

5. Which of the following dienes would you expect to be most stable?

a.

b.

c.

d.

e.

Short Answer

6. Circle any conjugated portions of the molecules or ions below.

Topic Test 1: Answers

1. **True.** The product 1,3-cyclohexadiene is conjugated, whereas the reactant 1,4-cyclohexadiene is not. Because the product is lower energy than the reactant, the reaction will be exergonic (energy yielding).

2. **False.** Although many conjugated molecules do appear to have alternating double and single bonds, there are also many conjugated systems that do not have that form.

3. **b.** Note that the pi bond between carbons 1 and 2 is in the same plane as the one between carbons 3 and 4. There is another pi bond between carbons 2 and 3 that is not part of the conjugated system but rather is perpendicular to it.

4. **a**

5. **b** This diene is conjugated and the others are not.

6.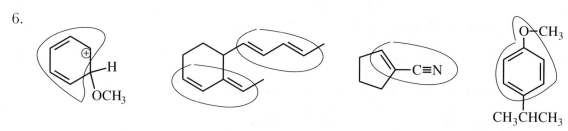

TOPIC 2: 1,2 AND 1,4 ADDITION, KINETIC VERSUS THERMODYNAMIC CONTROL

KEY POINTS

✓ What product(s) comes from reaction of HX to a conjugated diene?

✓ What is meant by 1,2 and 1,4 addition?

✓ What is the mechanism of 1,2 and 1,4 addition to a conjugated diene?

✓ What product is favored under thermodynamic or kinetic control?

✓ Under what conditions will 1,2 or 1,4 addition predominate?

Addition of one equivalent of HBr to 1,3-butadiene yields a mixture of two products in which the ratio changes depending on the reaction temperature. The product in which the H and Br end up on carbons 1 and 2 results from **1,2 addition**. Similarly, the **1,4 addition** product bears the H and Br on carbons 1 and 4, respectively.

$$CH_2=CH-CH=CH_2 \xrightarrow[\text{Ether}]{\text{HBr}} CH_3\underset{Br}{CH}-CH=CH_2 + CH_3CH=CH\underset{Br}{CH_2}$$

	1,2 Addition	1,4 Addition
0 °C	71%	29%
40 °C	15%	85%
	Product of kinetic control	Product of thermodynamic control

These results are explained by the mechanism. Initial capture of a proton by an end carbon (carbon 1) yields the most stable carbocation that is allylic and represented by two resonance forms. This allylic carbocation can be captured by the bromide nucleophile at either end of the 3-atom conjugated array. That leads to products in which the bromine is attached to carbon 2 or carbon 4. Attack by bromide at C(2) is faster, but attack at C(4) leads to a more stable product (a disubstituted alkene versus monosubstituted). The situation is depicted on the double energy versus reaction diagram below. The allylic carbocation intermediate (in the center) has two possible fates represented by paths to the left or right. The path on the left (1,2 addition) has lower activation energy and is therefore faster. This leads to the product of kinetic control, which will be favored by mild reaction conditions such as lower reaction temperature and short reaction time. The path to the right (1,4 addition) leads to a product that is lower in energy even though the activation barrier is greater. It is more difficult (slower) to make that product, but once formed it is more stable. This is the product of thermodynamic control, and it is favored by harsher reaction conditions such as higher temperature and longer reaction time.

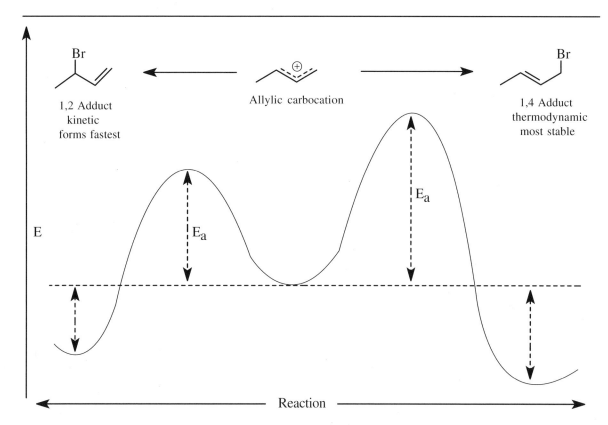

Electrophilic addition to conjugated dienes can generally occur by a 1,2 and/or 1,4 path. Note the numbers 1,2 and 1,4 refer to positions within the original diene and do not necessarily represent the numbering used in IUPAC nomenclature (although they are the same for the example above). Note also that it is not necessarily true that the kinetic product will be 1,2 and the thermodynamic product be 1,4 (as in HBr + 1,3-butadiene). Often the product that forms fastest is also the most stable.

Topic Test 2: 1,2 and 1,4 Addition, Kinetic Versus Thermodynamic Control

True/False

1. Cyclohexene can undergo 1,2 or 1,4 addition.
2. The product of thermodynamic control is the one that forms the fastest.

Multiple Choice

3. Which of the following could be obtained from a 1,2 addition to 1,3-pentadiene?

   a. [structure: CH₂(Br)–CH₂–CH=CH–CH₃]

   b. [structure: CH₃–CH=CH–CH(Br)–CH₃]

   c. [structure: CH₂=CH–CH₂–CH(Br)–CH₃]

   d. [structure: CH₂(Br)–CH=CH–CH=CH₂]

   e. None of these

4. Which of the following conditions favor formation of the product of kinetic control?
   a. Long reaction times
   b. Vigorous reaction conditions
   c. Lower temperature
   d. All of the above
   e. None of the above

Short Answer

5. Draw all reasonable resonance forms for the carbocation intermediate in the reaction of one equivalent of HBr to 5,5-dimethyl-1,3-cyclopentadiene.

6. Provide structural formulas for the products that are obtained from the reaction in problem 5 above. Label each product as a 1,2 or a 1,4 adduct.

Topic Test 2: Answers

1. **False.** Cyclohexene is a simple cyclic alkene. The dual paths of 1,2 and 1,4 addition are reserved for conjugated dienes.

2. **False.** The product of thermodynamic control is the one that is the most stable. (It can also be the one that forms fastest, but that is not a requirement.)

3. **b.** [structure: CH₃–CH=CH–CH(Br)–CH₃]

   Responses a and c each show the bromine on what would be carbon 1 of the diene system (not nomenclature 1) and that is not possible if the intermediate carbocation is allylic. Only b has a bromine on an allylic carbon and has the correct formula for HBr + pentadiene. Response d is not possible because it is still a diene and could not have resulted from addition to a diene.

4. **c.** Higher temperatures and longer reaction times favor the thermodynamic product because the energy and time required to get past the higher activation barrier is available. To favor the kinetic product we desire just enough energy and time to make it past the lowest activation energy barrier while depriving the reaction mixture of sufficient energy to go over the higher barrier.

5.

6.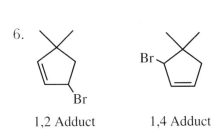

       1,2 Adduct   1,4 Adduct

TOPIC 3: THE DIELS-ALDER REACTION

KEY POINTS

✓ *What is the general form of a Diels-Alder reaction?*

✓ *What is the mechanism of a Diels-Alder reaction?*

✓ *What are the structural requirements for the Diels-Alder diene and dienophile?*

✓ *What product will result from a given diene + dienophile?*

✓ *What reactants will be needed to make a given Diels-Alder product?*

Many conjugated dienes can undergo cycloaddition reactions with other pi systems. The most common example is the Diels-Alder reaction between a **diene** and a **dienophile** ("diene-lover"). The simplest case (the "parent" reaction) is 1,3-butadiene + ethene, yielding cyclohexene as shown below. Note that the product has two more sigma bonds and two fewer pi bonds than the reactants. The product will always be a six-membered ring with a new pi bond between what were the two middle carbons of the original diene. Most Diels-Alder reactions can be viewed as extensions of the parent reaction.

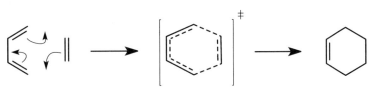

       Diene   Dienophile

The Diels-Alder reaction is also called a **[4 + 2] cycloaddition**. The mechanism is a concerted one-step process in which all relevant bonds are broken and formed simultaneously. This requires that the ends of the diene are pointed in the same direction relatively close to one another in a conformation called ***s-cis***.

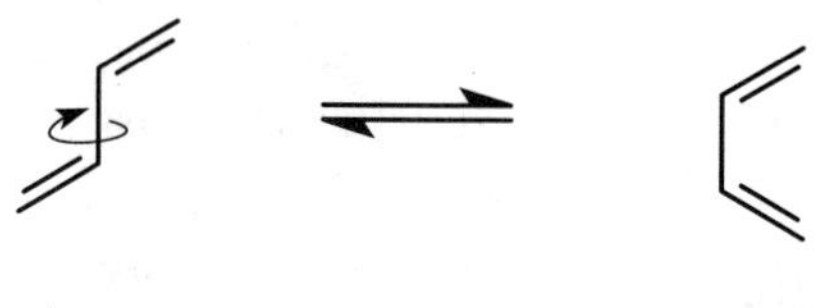

s-trans   *s-cis*

The concerted mechanism requires that the reaction to be **stereospecific**, that is, the stereochemistry of the reactants will be reflected in the products. In most cases, the best Diels-Alder dienophiles bear at least one electron-withdrawing group (—CO_2H, —CO_2R, —CN, —CHO, etc.) Some representative [4 + 2] cycloadditions are shown below.

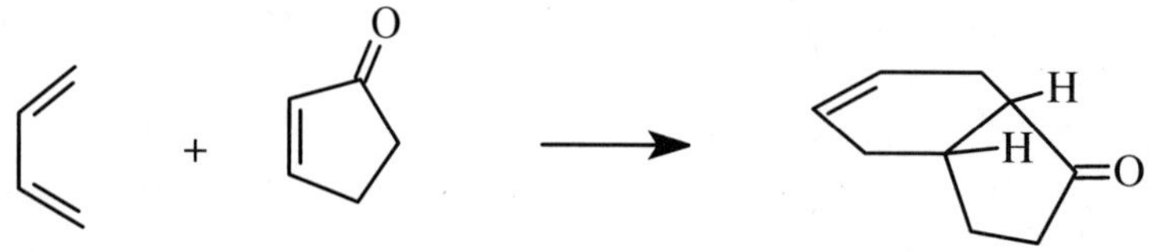

Rings are *cis* fused only

Ester groups are *cis* only

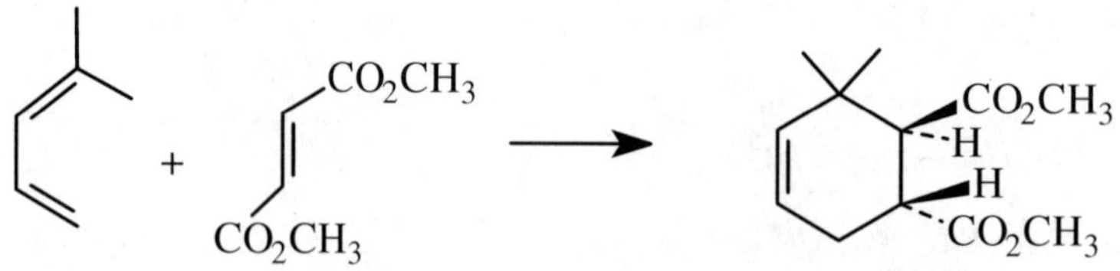

Ester groups are *trans* only

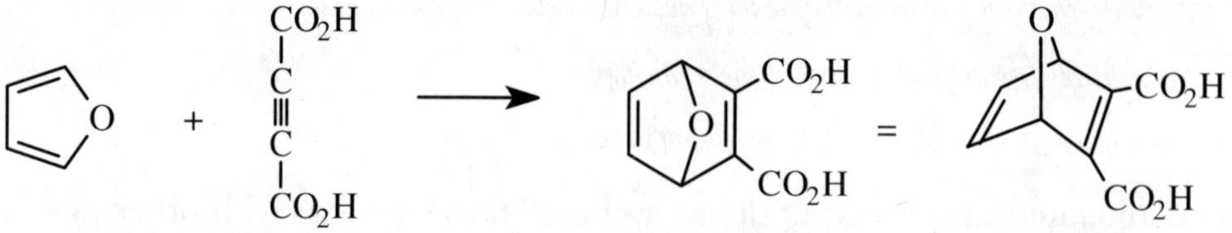

Topic Test 3: The Diels-Alder Reaction

True/False

1. Cyclopentene can be made by a Diels-Alder reaction.
2. The double bonds of 1,3-cyclopentadiene are "locked" into the *s-cis* arrangement.

Multiple Choice

3. Which of the following would be the most reactive in a normal Diels-Alder reaction with 1,3-butadiene?
   a. $H_2C=CH-CH=CH_2$
   b. $H_2C=CH-CH_2CH_3$
   c. $H_2C=CH-CH(CH_3)_2$

d. H₂C=CH—CH₂CH₂CH₂CH₃
e. H₂C=CH—CH=O

4. Which of the following statements is true about a Diels-Alder reaction?
   a. They are concerted.
   b. They are stereospecific.
   c. They are also called [4 + 2] cycloadditions.
   d. All of the above
   e. None of the above

Short Answer

5. Provide an unambiguous structural formula for the missing reaction product.

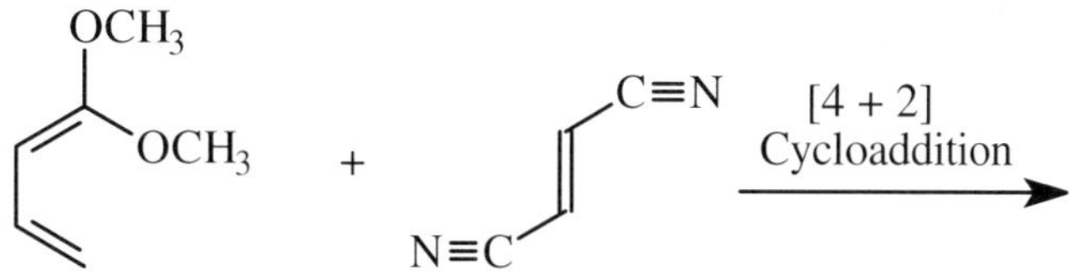

[diene with OCH₃, OCH₃ substituents] + [alkene with C≡N and N≡C] → [4 + 2] Cycloaddition

6. What starting materials would yield the following Diels-Alder product?

[4 + 2] → [bicyclic product with H and CO₂CH₂CH₃]

Topic Test 3: Answers

1. **False.** The product of a Diels-Alder reaction is a six-membered ring. Cyclopentene is too small to be made by this strategy.

2. **True.** This permanent *s-cis* orientation of the pi system makes molecules such as 1,3-cyclopentadiene react easily in Diels-Alder reactions.

3. **e.** This is the only alkene that has an electron withdrawing group (the aldehyde functionality) attached to it. b, c, and d are all simple alkyl-substituted ethenes, and a is 1,3-butadiene itself (the diene of the proposed reaction).

4. **d**

5. 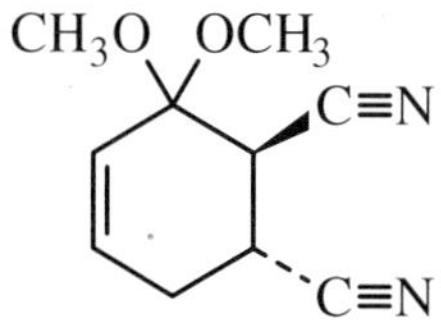 [cyclohexene ring with CH₃O, OCH₃, C≡N, C≡N substituents]

   Recall that the six-membered ring will have a "new" pi bond in the middle of what was formerly the diene. Note also that the cyano groups will be *trans* to one another on the ring. There are actually two enantiomers of this chiral product formed but only one is shown here.

6.  [1,3-cyclohexadiene] + H₂C=CHCO₂CH₂CH₃

Because there are several possible six-membered rings to consider, we can decide among them, recognizing that the "new" ring formed in the reaction must contain the pi bond. Recall also that the best dienophiles have electron-withdrawing groups on them.

APPLICATION

2-Methyl-1,3-butadiene is also known by its common name, **isoprene**. This conjugated diene is considered a building block for many natural products called **terpenes**. These contain multiples of five carbons (10, 15, 20, etc.) and obey the **isoprene rule**. According to this rule, terpenes have structures that appear to be assembled by "head to tail" joining of isoprene units. Oxygen-containing terpenes are sometimes called **terpenoids**. Some examples of terpenes and terpenoids are shown below with the isoprene units marked off.

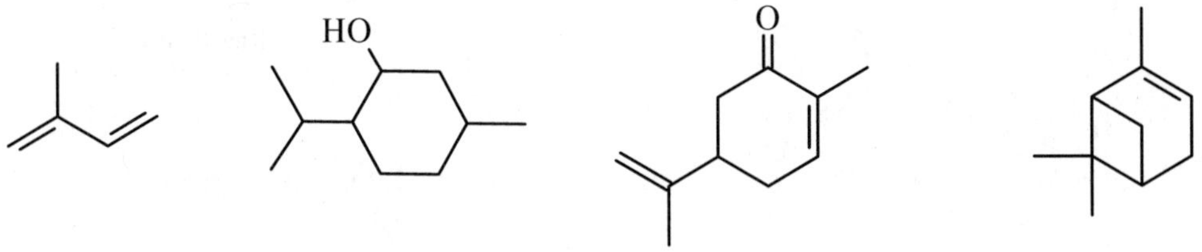

Isoprene   Menthol   Carvone   α-Pinene

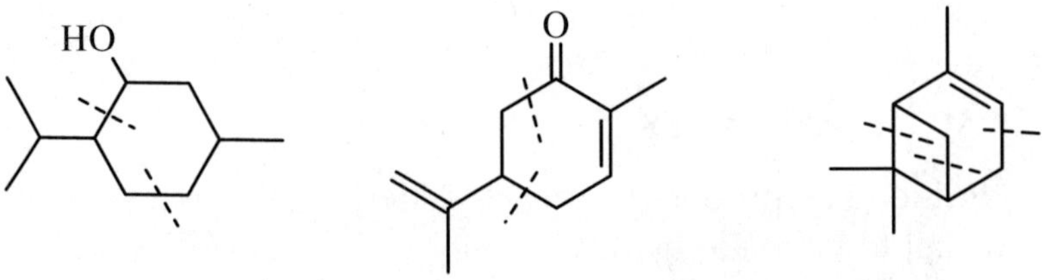

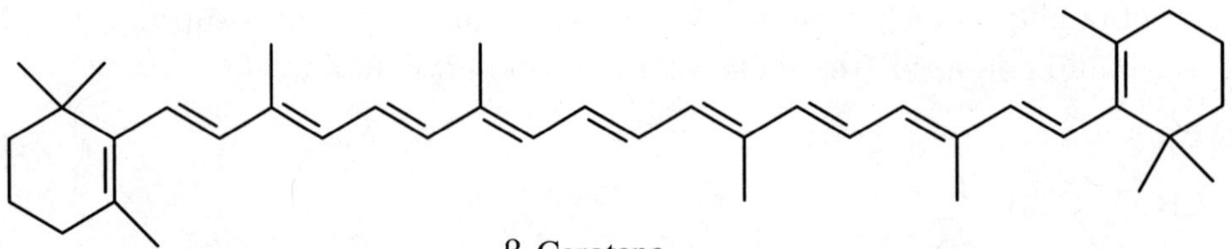

β-Carotene

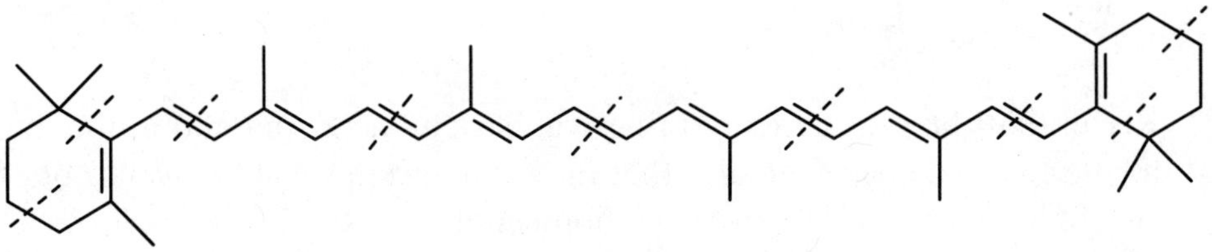

DEMONSTRATION PROBLEM

Show how the compound below could be prepared in one step from a Diels-Alder reaction. Identify the diene and the dienophile.

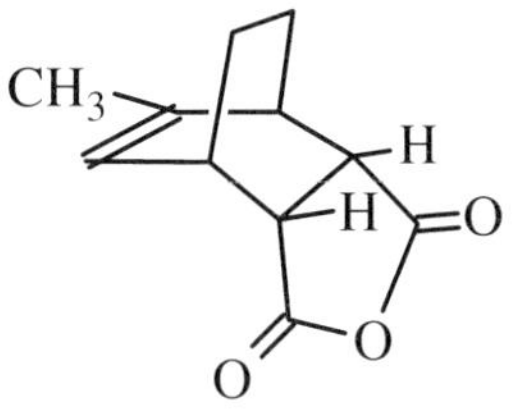

Solution

There are several rings in the product so we should first establish which resulted from the cycloaddition. A Diels-Alder reaction makes a six-membered ring with a pi bond in the middle of what was the diene. Recall also that normally the best dienophiles bear one or more electron-withdrawing groups (carbonyls or C≡N, etc.). Note also that because the product is tricyclic (three rings, i.e., it requires breaking three bonds to make an acyclic structure), two of the rings must have been in the reactants and the third resulted from the cycloaddition. Visualizing the reactants in a distorted geometry as shown below may be helpful, but the final answer should be expressed in a more spatially accurate format.

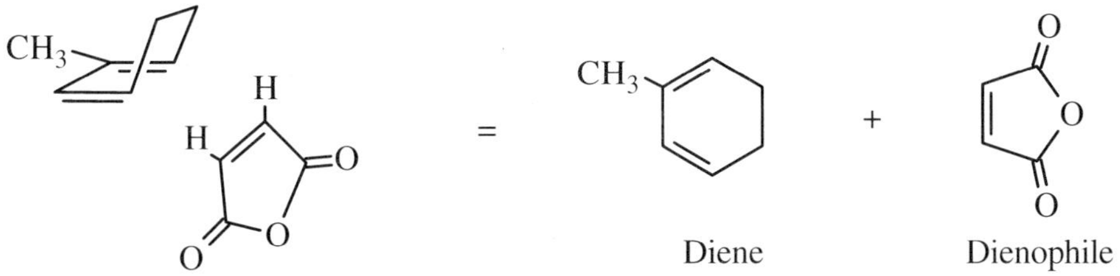

Chapter Test

True/False

1. Conjugated dienes are generally more stable than nonconjugated dienes.

2. The intermediate that leads to 1,2 or 1,4 addition of HX to a conjugated diene is an allylic carbocation.

3. Electron-withdrawing substituents such as CO_2R or CN will usually enhance the reactivity of a Diels-Alder dienophile.

4. The *s-cis* and *s-trans* conformers of 1,3-butadiene can be interconverted without breaking bonds.

5–10. Consider the hypothetical reaction curve below where some intermediate (I) has two possible fates leading to two different products (P1 and P2). For each of the following, indicate "P1" or "P2" as appropriate.

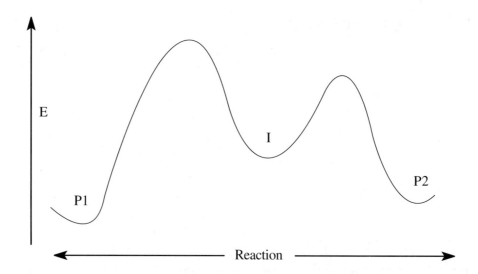

5. Forms fastest
6. Most stable
7. Favored by higher temperature and/or longer reaction time
8. Favored by lower temperature and/or shorter reaction time
9. Product of kinetic control
10. Product of thermodynamic control
11. Which of the following would react the fastest with a given dienophile?

   a.

   b.

   c.

   d.

   e. All of the above would react at the same rate.

12. Which of the following would react slowest with a given dienophile?

   a.

b.

c. [cyclopentane with exocyclic methylene and one ring double bond]

d. [cyclopentadiene]

e. All of the above would react at the same rate.

13. Which of the dienes below is the most stable?
   a. H$_2$C=CH—CH=CH—CH$_2$CH$_3$
   b. H$_2$C=CH—CH$_2$—CH=CH—CH$_3$
   c. H$_2$C=CHCH$_2$CH$_2$CH=CH$_2$
   d. H$_2$C=C=CHCH$_2$CH$_2$CH$_3$
   e. All of the above are isomers and therefore have the same stability.

Short Answer and Essay

14–15. Complete the following with unambiguous structural formulas for the missing organic compounds.

14. [diene] $\xrightarrow{[4+2]}$ [bicyclic diketone product with two CH$_3$ groups]

15. [2,3-dimethylbutadiene] + [CH$_2$=CH—CO$_2$H] $\xrightarrow{[4+2]}$

16–18. Circle the conjugated parts of the following.

16. [cyclohexenone with ethyl substituent]

17. [cyclohexadienyl cation with OCH$_3$, H, Cl substituents]

18. [cyclohexadiene with C≡N and C(=O)CH$_3$ substituents]

19. Although 1,3-cyclohexadiene is conjugated, addition of HX gives only one product regardless of temperature. Explain this observation.

Chapter Test 145

20. *Cis*- and *trans*-1,3-pentadiene react with a given dienophile at significantly different rates. Which one reacts faster? Offer an explanation for this observation.

Chapter Test: Answers

1. **T** 2. **T** 3. **T** 4. **T** 5. **P2** 6. **P1** 7. **P1** 8. **P2** 9. **P2** 10. **P1** 11. **d** 12. **c** 13. **a**

14.

15. [structure shown]

16. [structure shown]   17. [structure shown]   18. [structure shown]

19. Initial protonation of the diene yields a symmetrical allylic carbocation that can be attacked at either end by the halide nucleophile to yield 3-halocyclohexene. The same product results whether the mode of addition is 1,2 or 1,4.

20. The *trans* isomer reacts much faster than the *cis* because it is easier for the *trans* isomer to assume the required *s-cis* conformation.

Check Your Performance

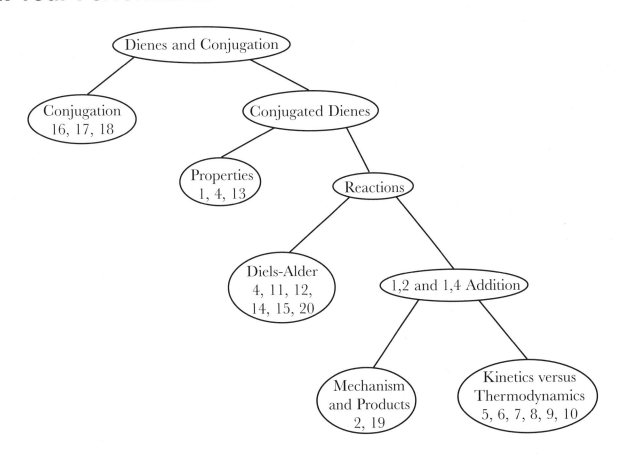

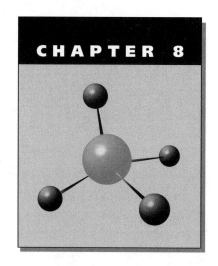

Aromatic Compounds

CHAPTER 8

At one time organic compounds were classified as "aromatic" based largely on their pleasant fragrances. As organic chemistry evolved, this term began to include compounds under this classification based on their structures and patterns of reactivity rather than their ability to stimulate an olfactory response. In this chapter we examine the structural criteria for aromaticity and look at some of the most important transformations of aromatic compounds.

ESSENTIAL BACKGROUND

- Resonance (Chapter 1)
- Conventions for writing mechanisms (Chapter 3)
- Electrophilic addition (Chapter 3)
- Nucleophiles and leaving groups (Chapter 6)

TOPIC 1: AROMATICITY, BENZENE, AND RESONANCE

KEY POINTS

✓ *What is meant by the term "aromatic"?*

✓ *What structural requirements define aromaticity?*

✓ *What is a Hückel number?*

✓ *What is the structure of benzene?*

✓ *What are some characteristic properties of aromatic compounds?*

To a modern organic chemist, the term **aromatic** has nothing to do with how a compound smells. Some aromatic compounds have no aroma at all. This term is used to describe compounds that are cyclic, planar, and completely conjugated around the ring and have a **Hückel number** of pi electrons. A Hückel number is given by $4n + 2$, where "n" is an integer (n = 0, 1, 2, 3, . . .), and therefore the Hückel numbers are 2, 6, 10, 14, ... A typical aromatic compound is benzene. Although a traditional Lewis-Kekule' representation of benzene would suggest it has alternating double and single bonds, experimentally it has been shown that all C—C bonds of benzene are the same length and strength. The apparent three pi bonds are not localized but rather are spread out over the entire ring. The C—C bonds are neither single nor double but rather are something between these extremes. This situation is shown below with resonance forms. Benzene and its derivatives are also represented with a circle signifying the six pi electrons of the ring.

Benzene and other aromatic compounds are unusually stable. The apparent pi bonds of benzene do not, in general, undergo addition reactions like those of alkenes (Chapter 3) but rather tend to undergo substitution reactions.

Topic Test 1: Aromaticity, Benzene, and Resonance

True/False

1. 1,3-Cyclobutadiene is aromatic.

2. 1,3,5-Cycloheptatriene is aromatic.

Multiple Choice

3. Which of the following ions is aromatic?

    a.   b.   c.

    d. All of the above
    e. None of the above

4. Which of the following heterocyclic compounds is aromatic?

    a.   b.   c.

    d. All of the above
    e. None of the above

Short Answer

5. List the structural requirements for a compound to be aromatic.

6. 7-Iodo-1,3,5-cycloheptatriene can lose an iodide ion to form a relatively stable carbocation with the formula $C_7H_7^+$. Show the structure of this cation and explain its stability.

Topic Test 1: Answers

1. **False.** Although 1,3-cyclobutadiene is cyclic and completely conjugated and can be represented by two equivalent resonance forms, it has four pi electrons and therefore does not conform to the $4n + 2$ rule and is not aromatic.

2. **False.** Although 1,3,5-cycloheptatriene is cyclic and has six pi electrons, it is not completely conjugated all the way around the ring. Carbon 7 is a saturated methylene, CH_2.

3. **b.** The cyclopentadienyl anion is aromatic. It is cyclic, completely conjugated around the ring, and has six pi electrons. Answers a and c each have four pi electrons.

4. **d.** In each of cases a, b, and c, the cyclic conjugated system contains six pi electrons. Knowing where the "nonbonded" or "lone pair" electrons reside is essential here.

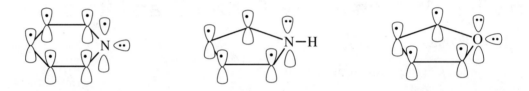

5. Aromatic molecules or ions must be cyclic, planar, and completely conjugated around the ring and must have a Hückel number $(4n + 2)$ of pi electrons $(2, 6, 10, 14, \ldots)$.

6. The stable ion is a cycloheptatrienyl cation, which is aromatic.

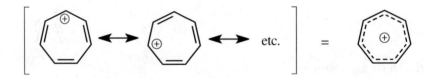

TOPIC 2: AROMATIC NOMENCLATURE

KEY POINTS

✓ *How are benzene derivatives named?*

✓ *What parent aromatic names are used to derive other names?*

✓ *What do the terms* ortho, meta, *and* para *mean with regard to substituents?*

✓ *How are benzene and related compounds numbered?*

Monosubstituted benzenes are named as such in many cases; however, there are some widely recognized **parent names** in which the single substituent on benzene gives the compound a different parent name.

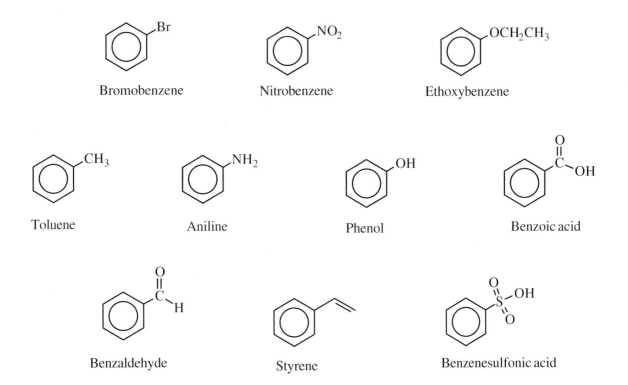

The carbon bearing the substituent in a monosubstituted benzene is understood to be carbon number 1. Any second substituent will have its location specified as 2 (***ortho***, *o*), 3 (***meta***, *m*), or 4 (***para***, *p*). If more than two substituents are attached to a benzene ring, the prefixes *o*, *m*, and *p* are not used and number locators are used exclusively. Substituents are listed in alphabetical order and the numbering is in the direction that gives the lowest numbers.

The aromatic substituent group C_6H_5 is called **phenyl** (not benzyl) and is often abbreviated Ph. A **benzyl** group is a phenyl attached to a methylene (i.e., —CH_2Ph).

Topic 2: Aromatic Nomenclature

1,2,3,4-Tetraphenyl-1,3-cyclopentadiene

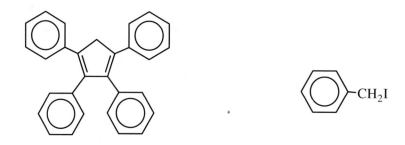

Benzyl iodide

Topic Test 2: Aromatic Nomenclature

Short Answer

1–3. Name the following.

Provide unambiguous structural formulas for the compounds named below.

4. *p*-Ethylbenzaldehyde
5. 2,4-Dibromostyrene
6. *m*-Dinitrobenzene

Topic Test 2: Answers

1. *ortho*-Iodotoluene (or *o*-iodotoluene or 2-iodotoluene)
2. 2-Phenyl-1,3-cycloheptadiene
3. 3,5-Dichlorophenol

4. CH₃CH₂—⟨C₆H₄⟩—C(=O)H
5. (2,4-dibromostyrene structure)
6. (1,3-dinitrobenzene structure)

TOPIC 3: GENERAL EAS REACTION AND SOME SPECIFIC EXAMPLES

KEY POINTS

✓ *What is the general mechanism of electrophilic aromatic substitution?*

✓ *What are some common examples of electrophilic aromatic substitution?*

The general mechanism for **electrophilic aromatic substitution** (EAS) involves attack of the ring's pi cloud on the electrophile to give a resonance-stabilized intermediate. Loss of a proton from the site of electrophilic attachment rearomatizes the ring.

Several common types of EAS reactions are shown in **Table 8.1**. The mechanisms for these reactions differ mostly in the steps leading up to the formation of the electrophile, E^+.

Table 8.1 Survey of Some EAS Reactions			
NAME	REAGENTS	E+ (EFFECTIVE)	PRODUCT
Chlorination	$Cl_2/FeCl_3$	(Cl^+)	Ar—Cl
Bromination	$Br_2/FeBr_3$	(Br^+)	Ar—Br
Nitration	HNO_3/H_2SO_4	NO_2^+	Ar—NO_2
Sulfonation	H_2SO_4/SO_3	HSO_3^+	Ar—SO_3H
Alkylation	R—X/$AlCl_3$	R^+ (R^+)	Ar—R
Acylation	RCOCl/$AlCl_3$	$RC\equiv O^+$	Ar—COR

Topic Test 3: General EAS Reaction and Some Specific Examples

Short Answer

Provide unambiguous structural formulas for the organic products that result from treatment of benzene with each of the following reagents and/or conditions.

1. $Cl_2/FeCl_3$

2. SO_3/H_2SO_4

3. $(CH_3)_3CCl/AlCl_3$

4. Draw all reasonable resonance forms for the intermediate that forms during the bromination of benzene.

Specify the reagents and/or conditions one could use to convert benzene into each of the following.

5. Nitrobenzene

6. Ph—C(=O)—CH_2CH_3

Topic Test 3: Answers

1. C₆H₅—Cl (chlorobenzene)
2. C₆H₅—SO₃H (benzenesulfonic acid)
3. C₆H₅—C(CH₃)₃ (tert-butylbenzene)

4. [Three resonance structures of the arenium ion intermediate with H and Cl on the sp³ carbon, positive charge delocalized around the ring]

5. HNO₃, H₂SO₄

6. CH₃CH₂C(=O)Cl, AlCl₃

TOPIC 4: SUBSTITUENT EFFECTS IN EAS REACTIONS

KEY POINTS

- ✓ *Will a given ring substituent speed up or slow down an EAS reaction?*
- ✓ *What are activating and deactivating groups?*
- ✓ *What regiochemistry results from an EAS reaction on a substituted benzene?*
- ✓ *What are ortho/para and meta-directing groups?*
- ✓ *How can one predict the effect of a particular substituent?*

A substituent on an aromatic ring can render that ring more or less reactive than benzene in EAS reactions. Those substituents that make the ring more reactive are called **activating groups** and those that make the ring less reactive are called **deactivating groups**. A substituent can also influence where an electrophile will attach to the ring. Some groups are called **ortho/para directors** because they tend to direct the electrophile to those positions. Other groups are known as **meta directors** because their presence results in predominately *meta* substitution. The activating/deactivating and o/p,m-directing characteristics are mostly predictable and can be generalized with a few rules.

Rule 1. Activating groups are usually also *o/p* directors and have at least one nonbonded electron pair on the atom attached directly to the ring.

$$-\ddot{N}H_2 \quad -\ddot{N}R_2 \quad -\ddot{O}H \quad -\ddot{O}R \quad -\ddot{O}CR(=O) \quad -\ddot{N}(H)CR(=O)$$

Rule 2. Deactivating groups are usually also *meta* directors and have the general form —W=Y, where an unsaturation (usually a pi bond) is in conjugation with the ring and Y is an electronegative atom.

154 Chapter 8 Aromatic Compounds

−NO₂	−SO₃R	−CO₂H	−COR	−CONR₂	−CN

$$-\overset{\oplus}{N}\overset{O}{\underset{O^{\ominus}}{\diagup}} \quad -\underset{O}{\overset{O}{\underset{\parallel}{S}}}-OR \quad -C\overset{O}{\diagdown}_{OH} \quad -C\overset{O}{\diagdown}_{R} \quad -C\overset{O}{\diagdown}_{NR_2} \quad -C\equiv N$$

Rule 3. Alkyl groups are mildly activating and are *o/p* directors but they do not fit the general form described in rule 1 above.

$$-CH_3 \quad -CH_2CH_3 \quad -CH(CH_3)_2 \quad -C(CH_3)_3$$

Rule 4. Halogens and nitroso are mixed effect substituents in that they are mildly deactivating but favor *ortho/para* substitution rather than *meta*.

$$-Cl \quad -Br \quad -I \quad -N=O$$

Topic Test 4: Substituent Effects in EAS Reactions

True/False

1. Monobromination of aniline yields mostly *meta*-bromoaniline.

2. Toluene undergoes most EAS reactions faster than benzaldehyde does.

Multiple Choice

3. Which compound below is likely to nitrate at the *meta* position?

   a. C₆H₅−OCCH₃ (with C=O)   b. C₆H₅−NHCCH₃ (with C=O)   c. C₆H₅−NO₂

   d. All of the above
   e. None of the above

4. Which compound below will likely undergo acylation the fastest?
   a. Benzene
   b. Ethylbenzene

   c. C₆H₅−CNH₂ (with C=O)   d. C₆H₅−C≡N   e. C₆H₅−O−CH(CH₃)₂

Short Answer

5–6. Show the structure(s) expected for the monosubstitution products of each reaction below.

5. [α-tetralone] $\xrightarrow{H_2SO_4,\ SO_3}$

6. [Ph-CH$_2$-C$_6$H$_4$-NO$_2$] $\xrightarrow[AlCl_3]{CH_3CCl=O}$

Topic Test 4: Answers

1. **False.** The NH$_2$ group on the aromatic ring of aniline is an *ortho/para* director so the monobromination reaction will likely yield a mixture of *ortho*- and *para*-bromoaniline.

2. **True.** The deactivating carbonyl conjugated with the ring in benzaldehyde slows down any EAS reaction. The mildly activating methyl group of toluene enhances the reactivity of that aromatic ring toward EAS.

3. **c.** Nitrobenzene bears a *meta*-directing substituent. The product of further nitration will likely be *m*-dinitrobenzene. Compounds shown in responses a and b both have lone pairs conjugated with the aromatic ring. Substituents of this form are *o/p* directors.

4. **e.** PhOCH(CH$_3$)$_2$ contains the most activated aromatic ring. Ethyl benzene is only mildly activated, and the compounds pictured in c and d are deactivated.

5. The aromatic ring bears two substituents (they happened to be connected to one another but no matter). An incoming electrophile will preferentially attach to a position *ortho* or *para* to the *o/p*-directing alkyl group and *meta* to the *meta*-directing nitro group.

[Structures: 5-sulfo-α-tetralone (SO$_3$H at position para to alkyl) + 7-sulfo-α-tetralone (HO$_3$S ortho to alkyl)]

6. The two rings of this starting material differ in reactivity. The ring bearing the nitro group is deactivated relative to the other ring that has only an alkyl substituent that is a mildly activating *o/p* director. *Para* substitution will likely predominate for steric reasons, but some *ortho* is also expected on that ring.

[Structure: CH$_3$C(=O)-C$_6$H$_4$-CH$_2$-C$_6$H$_4$-NO$_2$]

156 Chapter 8 Aromatic Compounds

TOPIC 5: NUCLEOPHILIC AROMATIC SUBSTITUTION

KEY POINTS

✓ *What products come from reactions of nucleophiles with aromatic compounds?*
✓ *What mechanism(s) are possible for nucleophilic aromatic substitution?*
✓ *What is benzyne?*

Some substitution reactions on aromatic substrates occur by other (non-EAS) mechanisms. If the starting material bears a suitable leaving group, nucleophilic substitution can take place by one of two mechanisms. Aromatics that bear one or more electron-withdrawing groups (deactivating groups for EAS) in positions *ortho/para* to the leaving group favor an **addition-elimination** mechanism.

Those aromatics that bear a leaving group but do not have electron-withdrawing substituents can react under more vigorous conditions via an **elimination-addition** pathway known as the **benzyne** mechanism. Note that the regiochemistry is sometimes uncertain for benzyne reactions.

Topic Test 5: Nucleophilic Aromatic Substitution

Multiple Choice

1. Which of the following is most likely to undergo a nucleophilic aromatic substitution reaction via the addition-elimination mechanism?
   a. Chlorobenzene
   b. *meta*-Dinitrobenzene
   c. 1-Chloro-2,4-dinitrobenzene
   d. All of the above
   e. None of the above

2. Which of the following will likely react with KOH/H$_2$O at elevated temperature and pressure to yield a single organic product via the benzyne pathway?
   a. Chlorobenzene
   b. *p*-Chlorotoluene

c. *m*-Chlorotoluene
d. *o*-Chlorotoluene
e. All of the above

Short Answer

3. What reagents and conditions might be used to covert bromobenzene into aniline?

4–6. Provide unambiguous structural formulas for the missing organic products.

4. 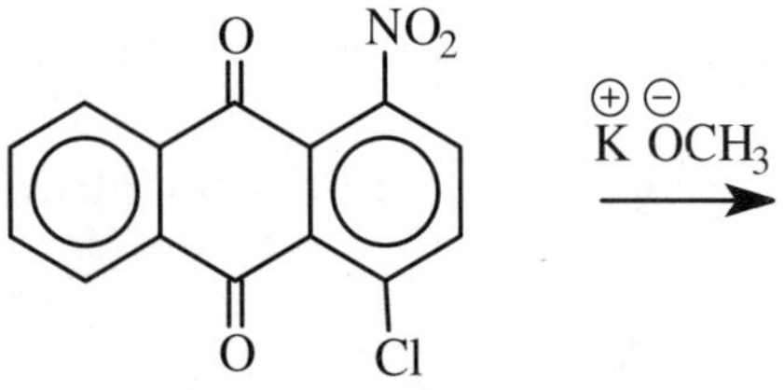 [anthraquinone with NO₂ and Cl substituents] + K⁺ ⁻OCH₃ →

5. 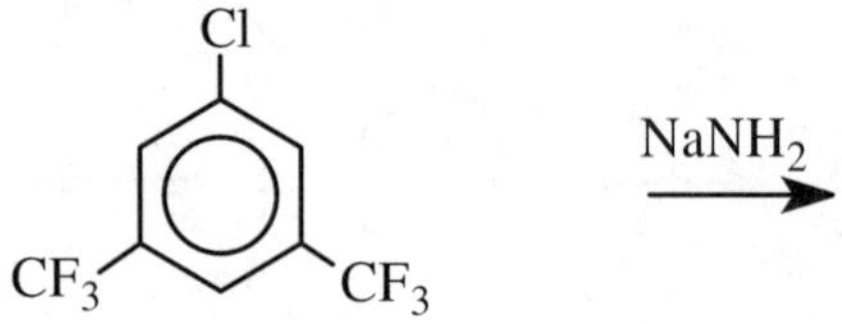 [benzene ring with Cl and two CF₃ groups] + NaNH₂ →

6. 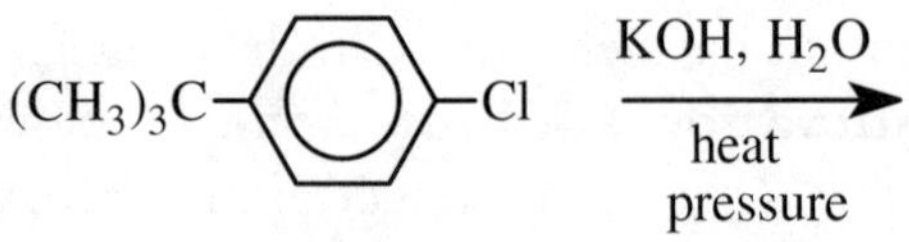 (CH₃)₃C—[benzene]—Cl $\xrightarrow{\text{KOH, H}_2\text{O, heat, pressure}}$

Topic Test 5: Answers

1. **c.** For the addition-elimination reaction mechanism to be feasible, the aromatic ring must bear a leaving group (chloride in this case) and electron-withdrawing groups on the ring that will stabilize the anionic intermediate.

2. **a.** The three chlorotoluene isomers would each yield more than one product via the benzyne mechanism

3. NaNH₂/NH₃ with heat and pressure (benzyne mechanism).

4. 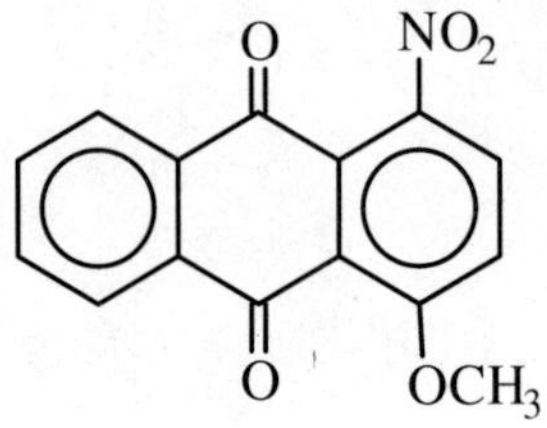 [anthraquinone with NO₂ and OCH₃ substituents]

5. 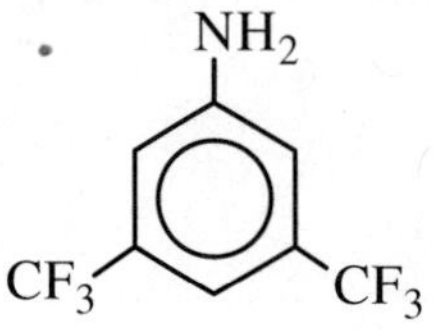 [benzene ring with NH₂ and two CF₃ groups]

6. 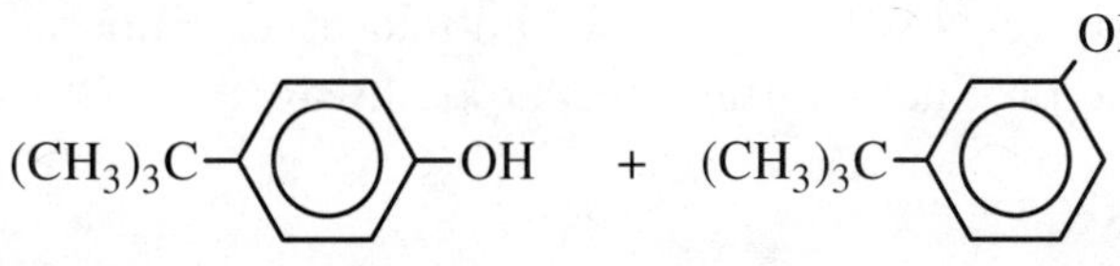 (CH₃)₃C—[benzene]—OH + (CH₃)₃C—[benzene with OH at meta position]

158   Chapter 8   Aromatic Compounds

TOPIC 6: OTHER REACTIONS OF AROMATIC COMPOUNDS

KEY POINTS

✓ *How can aromatic rings be reduced (hydrogenated)?*
✓ *What is benzylic bromination and what reagents will cause it?*
✓ *How can alkyl substituents on aromatic rings be oxidized?*

The pi systems of benzene and its derivatives are unreactive to the hydrogenation reactions used for alkenes and alkynes. Reduction of toluene to methylcyclohexane, for example, requires the use of a powerful catalyst and/or high pressure.

The alkyl side chains attached to aromatic rings often react at the position adjacent to the ring (called the **benzylic** position). Halogenation or oxidation at the benzylic position is generally possible provided there is at least one benzylic hydrogen atom in the starting material. Chlorination and bromination can be carried out under free radical conditions with molecular halogen or *N*-bromosuccinimide (NBS).

When alkyl groups with one or more benzylic hydrogen atoms are treated with hot aqueous permanganate, all carbons except benzylic carbon are cleaved off, and the resulting product is an aromatic carboxylic acid.

Topic Test 6: Other Reactions of Aromatic Compounds

True/False

1. Benzylic hydrogens are attached directly to the aromatic ring.
2. *t*-Butylbenzene can be oxidized to benzoic acid with $KMnO_4/H_2O$/heat.
3. The hydrogenation conditions used to reduce aromatic rings are usually more vigorous than those used for the hydrogenation of alkenes or alkynes.

Multiple Choice

4. What reagents and conditions could be used to convert *trans*-1,2-diphenylethene into 1,2-dicyclohexylethane?
   a. Excess H_2/Pd
   b. Excess H_2/Pt, pressure
   c. $KMnO_4/H_2O$/heat
   d. All of the above
   e. None of the above

Short Answer

5–6. Complete the following reactions with unambiguous structural formulas for the major organic products.

5.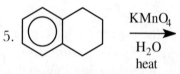

6.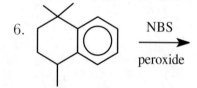

Topic Test 6: Answers

1. **False.** Benzylic hydrogens are those on the carbon attached directly to the ring.
2. **False.** The *t*-butyl group bears no benzylic hydrogens and therefore the permanganate cannot "bite into" the molecule. Recall that oxidations of alkyl side chains require benzylic hydrogens.
3. **True.** Alkenes and alkynes will usually hydrogenate at room temperature with a Pd catalyst, but aromatic rings normally require Pt or Rh catalysis and high pressure.
4. **b.** These hydrogenation conditions are strong enough to reduce the aromatic rings and therefore will also reduce the alkene.

5. [structure: benzene with two ortho CO₂H groups]

6. [structure: 1,1,4,4-tetramethyltetralin with Br at 4-position]

APPLICATION

The aromatic compound 2,4,6-trinitrotoluene is perhaps better known by its initials "TNT." It is a potent explosive once widely used in mining and demolition. It is a standard for comparison when expressing the explosive power of bombs. For example, a hydrogen bomb can have the destructive force equivalent to 10 million tons (10 megatons) of TNT.

[structure of TNT: toluene with NO₂ groups at 2, 4, and 6 positions]   TNT

DEMONSTRATION PROBLEM

Show the reagents and/or conditions one could use to prepare the compound below from toluene. More than one step may be required.

[structure: benzene with Cl and CO₂H in meta positions]

Solution

The carboxylic acid group in the product must have resulted from oxidation of the methyl on toluene using $KMnO_4$ and H_3O^+. The bromine was placed on the ring using Cl_2 and $FeCl_3$. Note that the order these reagents are applied is critical because different regiochemistry results if the steps are reversed.

$$\text{PhCH}_3 \xrightarrow[\text{H}_2\text{O, heat}]{\text{KMnO}_4} \text{PhCO}_2\text{H} \xrightarrow[\text{FeCl}_3]{\text{Cl}_2} \text{3-Cl-C}_6\text{H}_4\text{-CO}_2\text{H}$$

$$\text{PhCH}_3 \xrightarrow[\text{FeCl}_3]{\text{Cl}_2} \text{4-Cl-C}_6\text{H}_4\text{-CH}_3 \xrightarrow[\text{H}_2\text{O, heat}]{\text{KMnO}_4} \text{4-Cl-C}_6\text{H}_4\text{-CO}_2\text{H} \quad (+\ ortho)$$

Chapter Test

True/False

1. 1,3-Cyclohexadiene is aromatic.
2. Phenyl is an aromatic alcohol.
3. Aromatic compounds do not generally undergo addition reactions.
4. The formula for styrene is C₈H₈.
5. Benzene has no benzylic hydrogens.

Multiple Choice

6. Which of the following is a requirement for a compound to be aromatic?
   a. A strong odor
   b. Complete conjugation around a ring
   c. Must be benzene or a substituted benzene
   d. All of the above
   e. None of the above

7. Which step(s) below will convert toluene to *m*-chlorobenzoic acid?
   a. Water and Cl₂
   b. Cl₂/FeCl₃ followed by hot aqueous permanganate
   c. Hot aqueous permanganate and then Cl₂/FeCl₃
   d. All of the above
   e. None of the above

8. What step(s) below will convert benzene to *p*-bromobenzenesulfonic acid?
   a. H₂SO₄/HNO₃ and then Br₂/FeBr₃
   b. Br₂/FeBr₃ and then H₂SO₄/HNO₃
   c. H₂SO₄/SO₃ and then Br₂/FeBr₃
   d. Br₂/FeBr₃ and then H₂SO₄/SO₃
   e. None of the above

9. Which compound below will undergo EAS reactions the fastest?

   a. b. c.

   d. 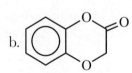 e.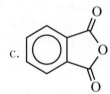

10. The most likely product from the reaction of toluene with NBS and peroxide is
    a. *ortho*-bromotoluene
    b. *meta*-bromotoluene

162   Chapter 8   Aromatic Compounds

c. *para*-bromotoluene
d. 2,4,6-tribromotoluene
e. benzyl bromide (PhCH₂Br)

Short Answer

11–16. Provide unambiguous structural formulas for the missing organic products.

11. 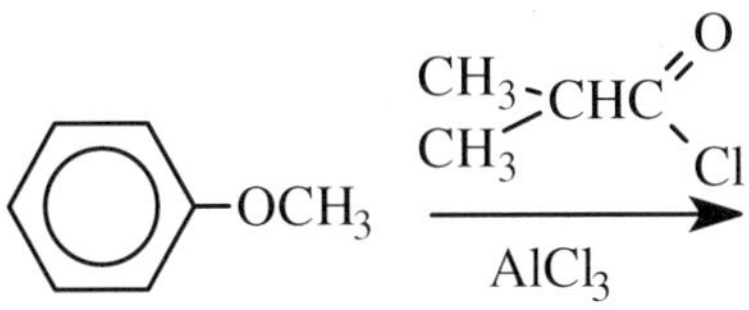 PhOCH₃ + (CH₃)₂CHC(=O)Cl →(AlCl₃)

12. 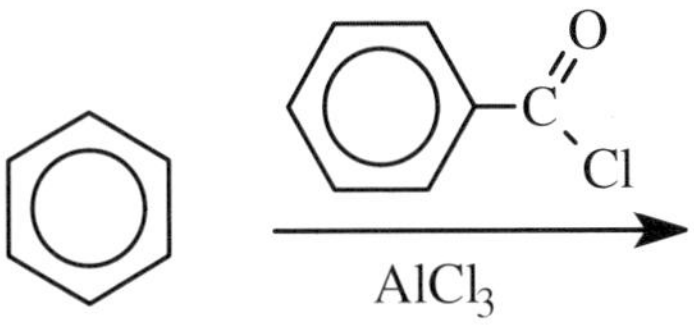 PhH + PhC(=O)Cl →(AlCl₃)

13. 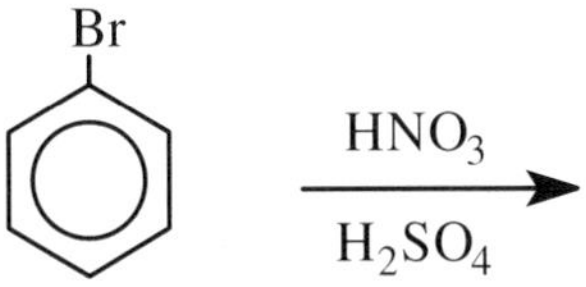 bromobenzene + HNO₃ / H₂SO₄ →

14. 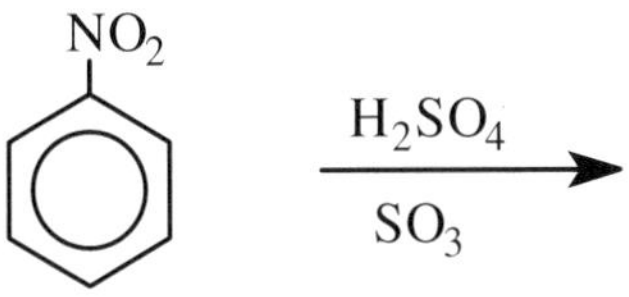 nitrobenzene + H₂SO₄ / SO₃ →

15. 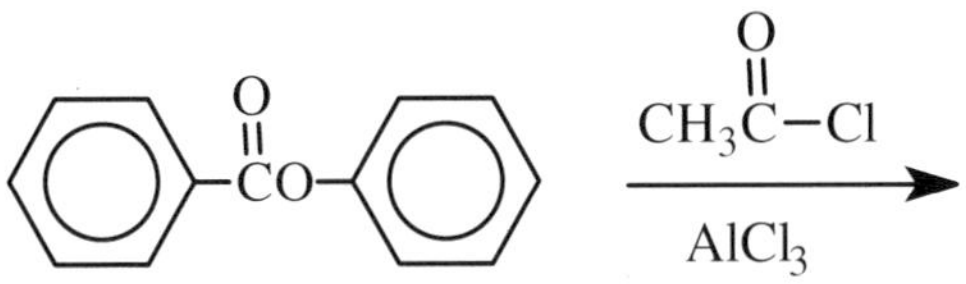 PhC(=O)O-Ph + CH₃C(=O)-Cl →(AlCl₃)

16. 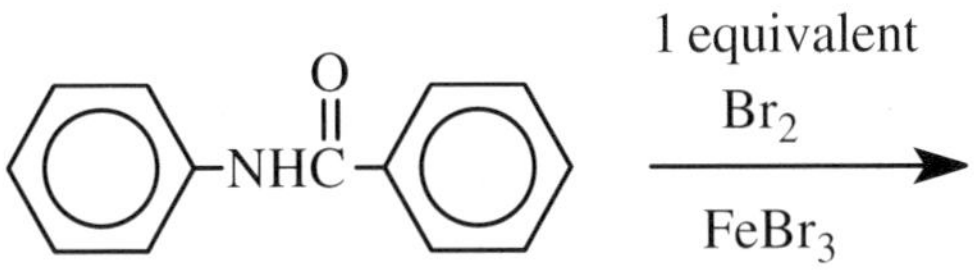 Ph-NHC(=O)-Ph + 1 equivalent Br₂ →(FeBr₃)

17. Sketch an orbital overlap picture of the pi system for the aromatic heterocyclic compound below. Indicate the positions of all pi and nonbonded electrons and specify the hybridization of all ring atoms.

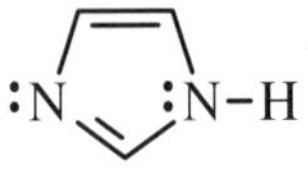

    (imidazole: :N, :N–H)

18. The *ortho*- and *para*-bromonitrobenzene each undergo nucleophilic aromatic substitution reactions with hydroxide at 130°C, yet *meta*-bromonitrobenzene is inert to these conditions. Explain this observation. (Hint: Consider resonance and the key mechanistic intermediates in each case.)

19. Show how one could synthesize *p*-nitrobenzoic acid from benzene or toluene and any other needed reagents. More than one step may be required.

20. Show the reagents and/or conditions one could use to carry out the transformation below. More than one step may be required.

Chapter Test: Answers

1. **F** 2. **F** 3. **T** 4. **T** 5. **T** 6. **b** 7. **c** 8. **d** 9. **b** 10. **e**

11. CH$_3$-CHC(=O)-C$_6$H$_4$-OCH$_3$ with (CH$_3$)$_2$CH- group (+ *ortho*)

12. C$_6$H$_5$-C(=O)-C$_6$H$_5$

13. Br-C$_6$H$_4$-NO$_2$ (+ *ortho*)

14. NO$_2$-C$_6$H$_4$-SO$_3$H

15. C$_6$H$_5$-C(=O)-O-C$_6$H$_4$-C(=O)CH$_3$ (+ *ortho* on same ring)

16. Br-C$_6$H$_4$-NHC(=O)-C$_6$H$_5$ (+ *ortho* on same ring)

17. All ring atoms sp^2 hybridized

18. The intermediates that result from nucleophilic attack on C(1) are resonance-stabilized anions. The *ortho* and *para* isomers have resonance forms in which the negative charge is stabilized by the strong electron-withdrawing nitro group. The nitro on the intermediate from the *meta* isomer is not able to delocalize the negative charge.

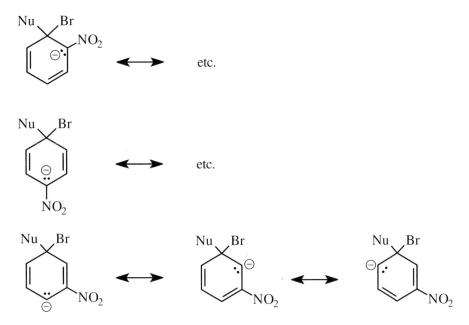

19. Treat toluene with HNO_3/H_2SO_4 and then hot aqueous permanganate.

20. $AlCl_3$ (intramolecular acylation) and chlorinate with $Cl_2/FeCl_3$.

Check Your Performance

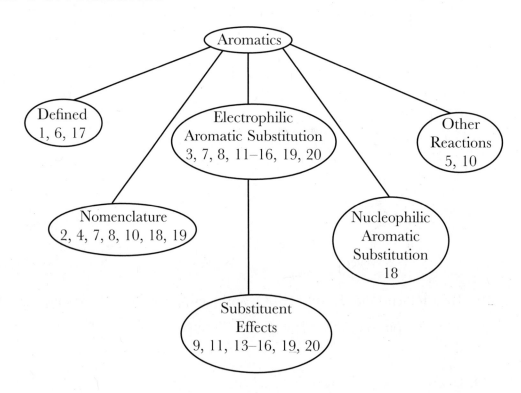

Chapter 9: Spectroscopy

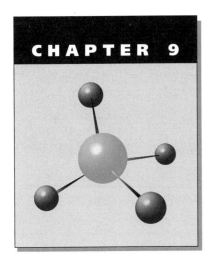

How does one know what products come from a given reaction? It is usually not possible to tell what the structural formula for a compound is by casual inspection of some container of it. How would one know, for example, whether or not the addition of HBr to $CH_3CH{=\!\!=}CH_2$ yielded the Markovnikov product? In this chapter we address this issue and survey some instrumental techniques chemists use to identify organic compounds.

ESSENTIAL BACKGROUND

- Resonance (Chapter 1)
- Functional groups (Chapter 2)
- Isomers (Chapters 2 and 5)
- Degree of unsaturation (Chapter 3)
- Relative stability of carbocations (Chapter 3)
- Relative stability of free radicals (Chapters 2 and 3)

TOPIC 1: MASS SPECTROMETRY

KEY POINTS

✓ *What takes place at the microscopic level when a mass spectrum is measured?*

✓ *What is a molecular ion?*

✓ *What is a base peak in a mass spectrum?*

✓ *What can be learned about a compound by examination of its mass spectrum?*

Most **mass spectra** are acquired using an instrument that places a small amount of gaseous sample in an evacuated chamber where it is bombarded by high-energy electrons. If one of these electrons strikes a molecule of sample with sufficient energy, one of the electrons from the sample molecule is dislodged, leaving behind a positively charged odd-electron species called a **radical cation**. Because the mass of the ejected electron is negligible, the mass and molecular formula for this radical cation are essentially the same as the original sample molecule. This radical cation is usually designated as M^+ and called the **molecular ion**. The charged particles are accelerated by a magnetic field toward a detector that records the impact and registers the mass. The detection equipment of a mass spectrometer is designed to detect ions but not uncharged species. A molecular ion may be so unstable that it decomposes before being detected. When a radical cation fragments, the resultant particles will be radicals and cations. Favorable fragmentation will result in the most stable cations and radicals. The intensity of a peak in the mass spectrum is proportional to the number of ions of that mass reaching the detector. The most intense (tallest) peak of a mass spectrum is called the **base peak**. In some cases the molecular ion is also the base peak. A mass spectrum usually allows one to determine the molecular mass of a compound (and therefore possible molecular formulas) and fragmentation pathways. Examining the spectrum for intensities and masses of molecular and fragment ions can help identify some molecules.

Topic Test 1: Mass Spectroscopy

True/False

1. The tallest (most intense) peak in a mass spectrum is the molecular ion.
2. The molecular ion for benzene will likely appear in the mass spectrum at m/z = 78.
3. The base peak of a mass spectrum is usually abbreviated as M^+.
4. Compounds with the same molecular formula will also produce the same mass spectrum.

Multiple Choice

5. Which pair below will most likely give the same molecular ion?
   a. Hexane and cyclohexane
   b. Cyclohexane and cyclohexene
   c. Cyclohexane and *trans*-2-hexene
   d. All of the above
   e. None of the above

6. Which compound will most likely show a strong peak at M-15 in its mass spectrum?
   a. Cycloheptane
   b. 1,1-Dimethylcyclopentane
   c. Biphenyl (Ph-Ph)
   d. All of the above
   e. None of the above

Short Answer

7. The mass spectrum of an unidentified hydrocarbon shows a $M^+ = 80$. Determine the formula and propose a possible structure.

8. Which compound below is most likely to show a large mass spectral peak at m/z = 69? Explain your choice.

$$(CH_3)_2C=CHCH_2CH_3 \quad (CH_3)_2C=C(CH_3)_2$$

Topic Test 1: Answers

1. **False.** The most intense peak in a mass spectrum is called the base peak. Although the molecular ion is sometimes also the base peak, that is not always the case.
2. **True.** The formula for benzene is C_6H_6 with a molecular weight (MW) of 78.

3. **False.** The abbreviation M⁺ stands for the molecular ion.

4. **False.** Isomers will have the same M⁺ but can, and often do, fragment differently and therefore have different mass spectra.

5. **c.** To have the same M⁺ the compounds must be isomers. Both these compounds have the formula C_6H_{12}.

6. **b.** This is the only compound listed that has methyl appendages. If one of the methyl groups is lost from the M⁺, the resulting fragment ion will have a mass that is 15 amu lighter and will be tertiary.

7. A hydrocarbon with an MW = 80 must be C_6H_8. Any compound with that formula is a possible correct answer: 1,3-cyclohexadiene or 1,3,5-hexatriene, etc.

8. $(CH_3)_2C=CHCH_2CH_3$ is most likely to show m/z = 69. Both compounds have MW = 84. A peak at 69 corresponds to loss of methyl (M-15). Loss of methyl from the molecular ion of $(CH_3)_2C=C(CH_3)_2$ would yield the relatively unstable vinylic carbocation. Loss of methyl from the molecular ion of $(CH_3)_2C=CHCH_2CH_3$ (i.e., methyl that is carbon 5 in 2-methyl-2-pentene) produces a resonance-stabilized allylic carbocation.

TOPIC 2: ULTRAVIOLET VISIBLE

KEY POINTS

✓ *What occurs when a molecule absorbs ultraviolet-visible (UV-vis) light?*

✓ *What kind of molecules will absorb UV-vis light?*

✓ *What can be learned about a compound by examining its UV-vis spectrum?*

UV-vis spectroscopy measures the transition of electrons from occupied to unoccupied molecular orbitals. The amount of energy required for this transition depends on the energy gap between the two orbitals. The kinds of transitions that fall within the usual UV-vis range are n → π* and π → π*. Conjugation lowers the ΔE between orbitals, and normally compounds that absorb UV-vis light of longer wavelengths than 210 nm have such conjugation. In general, more conjugation causes a longer wavelength of absorption. Conjugation can include both pi bonds and non-bonded electron pairs. Systems with similar conjugation tend to have similar UV-vis spectral characteristics. A **wavelength of maximum absorption** is designated by the symbol λ_{max}. Some representative UV-vis data are shown in **Table 9.1**.

Table 9.1 Some Ultraviolet-Visible Absorption Maxima	
COMPOUND	λ_{MAX} (NM)
R—CH=CH—R	~165
1,3-butadiene	217
1,3,5-hexatriene	258
1,3,5,7-octatetraene	290
2-methyl 1,3-butadiene	220
3-buten-2-one	219
Benzene	204, 254
Phenol	210, 270

Topic Test 2: Ultraviolet Visible

True/False

1. Cyclohexene absorbs UV-vis light at a wavelength longer than 210 nm.
2. *trans*-1,3-Pentadiene has a longer λ_{max} than 1,4-pentadiene does.
3. Methyl vinyl ether ($CH_3OCH=CH_2$) has a longer λ_{max} than $CH_3CH=CH_2$ does.

Multiple Choice

4. Which compound below will have the longest λ_{max}?
   a. Cyclohexane
   b. Cyclohexene
   c. Benzene
   d. Toluene
   e. Nitrobenzene

5. Which of the following is *not* an ultraviolet transition normally observed above 210 nm?
   a. n-π*
   b. π-π*
   c. σ-σ*
   d. All of the above
   e. None of the above

Short Answer

6. An unidentified compound with the formula C_6H_8 was found to have no UV-vis absorption above 200 nm. Propose a possible structure.

Topic Test 2: Answers

1. **False.** Cyclohexene has only a single pi bond and is not conjugated.
2. **True.** 1,3-Pentadiene is conjugated and 1,4-pentadiene is not.
3. **True.** The lone pairs of electrons on the oxygen are part of the extended pi system.
4. **e.** Besides the aromatic ring, the nitro group is also part of the conjugation.
5. **c.** This transition will only occur at very high energies that correspond to wavelengths much shorter than 210 nm.
6. There are many possible correct answers. Any C_6 hydrocarbon with three degrees of unsaturation and no conjugation is a reasonable choice (e.g., 1,4-cyclohexadiene or 3-vinylcyclobutene or cyclobutylethyne, etc.).

TOPIC 3: INFRARED SPECTROSCOPY

KEY POINTS

✓ *What occurs when a molecule absorbs infrared (IR) light?*
✓ *What can be learned about a compound from its IR spectrum?*
✓ *What are some important regions in the IR spectrum?*

To a first approximation, bonds connecting atoms can be viewed as springs connecting two masses. These "springs" bend and stretch with energies that depend on the masses of the attached atoms or groups and the inherent strength of the spring. The stretching and bending frequencies of the bonds are expressed in wavenumbers (cm^{-1}) and can be measured with IR spectroscopy. A typical IR spectrum extends from about 4000 cm^{-1} at left to about 400 cm^{-1} at right. When a molecule absorbs IR energy, the bonds are set into vibration. The precise position of the absorption frequency within the spectrum provides information on what "kinds" of bonds or functional groups are present. Some general regions of the spectrum and the types of vibrations to which they correspond are represented below along with a list of some common IR peak positions.

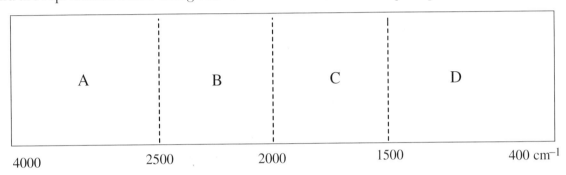

A: Stretching of single bonds to hydrogen (C—H, O—H, N—H)

B: Triple-bond stretch (C≡C, C≡N)

C: Double-bond stretch (C=C, C=O, C=N)

D: "Fingerprint region," stretching single bonds not to H, bond bending, harmonics.

Many of the most common and diagnostic bonds are shown in **Table 9.2**.

Table 9.2 Some Common IR Absorption Bands	
FUNCTIONAL GROUP	**APPROXIMATE BAND POSITION (CM^{-1})**
C—H	2850–2960
=C—H	3020–3100
C≡C—H	3300
Aromatic-H	3030
O—H	3400–3640 (alcohol), 2500–3100 (carboxylic acid)
N—H	3310–3500
C≡C	2100–2260
C≡N	2210–2260
C=O	1670–1780
NO$_2$	near 1350 and 1530
Aromatic C—C	near 1500 and 1600
C—O	1050–1150
C—Cl	600–800
C—Br	500–600
C—I	500

Topic Test 3: Infrared Spectroscopy

True/False

1. Cyclohexane and 1-hexene are isomers and therefore have nearly identical IR spectra.
2. Cyclohexane and cycloheptane are not isomers, yet they have nearly identical IR spectra.
3. Most bond bending occurs with energies that correspond to the fingerprint region.

Multiple Choice

4. The region of the IR spectrum where double bonds stretch is
   a. 4000–2500 cm^{-1}.
   b. 2500–2000 cm^{-1}.
   c. 2000–1500 cm^{-1}.
   d. the fingerprint region.
   e. None of the above

5. Which class of compounds below shows a strong IR absorption near 1700 cm^{-1}.
   a. Alcohol
   b. Ether
   c. Alkyne
   d. Aldehyde
   e. None of the above

Short Answer

6. Briefly explain how IR spectroscopy could distinguish between 1-pentyne and 2-pentyne.
7. Briefly explain how IR spectroscopy could distinguish between cyclohexane and benzene.

Topic Test 3: Answers

1. **False.** Although they are isomers, these compounds have different kinds of bonds and will therefore show different peaks in their IR spectra.
2. **True.** Both these compounds are medium-sized cyclic alkanes with similar bonds. They will therefore have similar absorption bands in their IR spectra.
3. **True.** The region to the right of 1500 wavenumbers (the fingerprint region) is where the bond-bending energies are found.
4. **c**
5. **d.** This is the only carbonyl-containing class in the list, and a peak near 1700 wavenumbers is likely a carbonyl.
6. The terminal alkyne 1-pentyne has a hydrogen atom bound directly to an sp-hybridized carbon. That bond stretches near 3300 cm^{-1}. The internal alkyne 2-pentyne will not have an IR signal in this region.

7. Cyclohexane is a cycloalkane that will show sp³ C—H stretch in the 2900-cm⁻¹ region. Benzene has only aromatic C—H stretches that appear near 3030 cm⁻¹. Benzene will also show the aromatic C—C stretches near 1500 and 1600 cm⁻¹.

TOPIC 4: NUCLEAR MAGNETIC RESONANCE AND ¹H-NMR

KEY POINTS

✓ *What kinds of nuclei does nuclear magnetic resonance (NMR) detect?*

✓ *What occurs when NMR is being measured?*

✓ *How does proton NMR reveal the number of different hydrogen environments?*

✓ *What is meant by "chemical shift" and what does it reveal?*

✓ *How can one determine the number of hydrogens in each environment?*

✓ *What information is contained in ¹H-NMR peaks that appear as multiplets?*

The **nuclear magnetic resonance** (NMR) phenomenon involves certain nuclei that can be "flipped" between energy states by radio frequency in the presence of an applied magnetic field. Nuclei of atoms that have an odd atomic number and/or odd atomic mass have a nuclear "spin." Among such nuclei are ¹H, ¹³C, ¹⁵N, ¹⁹F, and ³¹P. Spinning nuclei can be aligned with or against an applied magnetic field. Alignment with the applied magnetic field (parallel) is a lower energy state than alignment against the applied magnetic field (antiparallel). Nuclei can be stimulated to flip from the lower to the higher energy state when the correct amount of energy is applied. When that occurs, the nucleus is said to be "in resonance." The absorption of energy to bring a nucleus into resonance can be recorded in the form of an NMR spectrum. The energy required to induce this nuclear resonance corresponds to the long wavelengths and is normally in the range of radio waves. The precise energy required will depend on several factors, including the type of nucleus and the environment of that nucleus.

The most common kind of NMR spectroscopy is proton NMR (¹H-NMR). In general, a ¹H-NMR spectrum will help to answer four important questions about the structure of the compound:

Question 1: How many different ¹H environments are present? This is found through peak counting. In theory there is a different peak for each chemically different hydrogen environment within a molecule. Accidental superimposition of peaks is common so the actual number of peaks observed is often fewer than the number of ¹H environments.

Question 2: What is the nature of each hydrogen environment? The "chemical shift" or the position of the peak within the spectrum indicates this. Proton NMR shifts are reported in **parts per million** (ppm or δ) **downfield** (to the left) of teramethylsilane (**TMS**), which is assigned the value of zero. Generally, protons near electronegative substituents are shifted downfield to a larger extent. Sometimes proton shifts are affected strongly by the presence of multiple bonds or aromatic rings. Some rough ranges within which specific hydrogen types can be expected are listed in **Table 9.3**.

Table 9.3 Some Proton NMR Chemical Shift Ranges

Proton Environment and General Structure		Approximate Chemical Shift (δ)
Alkane-like		0–1.5
On sp³-hybridized carbons adjacent to unsaturation (Bezylic, Allylic, alpha to C=O)	=C-C-H	1.5–2.5
On carbon bearing electronegative atom (Z = halogen, O, N)	—CHZ—	2.5–4.5
Vinylic	C=C—H	4.5–6.5
Aromatic	Ar—H	6.5–8.0
Aldehydic	RCH=O	near 10
Carboxylic acid	RCO$_2$H	11–13

Question 3: How many protons of each type are present? This information appears in relative terms and is given by the relative areas under the peaks. Most NMR spectrometers are equipped with an integrator that converts peak areas into digitized numbers or altitudes for comparison. The relative areas of the peaks reflect the relative numbers of protons producing them.

Question 4: How many neighboring protons are adjacent to each proton type present? This is deduced from the **splitting pattern** caused by the **coupling** of nearby protons. In general, a proton NMR signal coupled to n equivalent nuclei will be split into n + 1 lines. If the neighbors are not strictly equivalent but coincidentally couple by similar amounts, the n + 1 relationship can still hold, but this is not always the case. More complex and non-first-order coupling patterns sometimes appear, especially when the differences between the chemical shifts of the nuclei are not much larger than the magnitude of the coupling between them. Some common coupling situations and patterns are given in **Table 9.4**.

Table 9.4 Some Common Bonding Arrangements and Proton NMR Coupling Patterns

Multiplet Name	Relative Line Areas	Possible Structures (G and X bear no H)
singlet		G—CH$_3$
doublet	1:1	CH$_3$—CHX$_2$ (CH$_3$)$_2$CH—G
triplet	1:2:1	CH$_3$—CH$_2$—G
quartet	1:3:3:1	CH$_3$—CH$_2$—G
quintet	1:4:6:4:1	(X—CH$_2$)$_2$CH—G
septet	1:6:15:20:15:6:1	(CH$_3$)$_2$CH—G

There are several common patterns of shift, integration, and splitting that you should recognize, including isolated methyl, ethyl, isopropyl, *t*-butyl, *p*-aromatic, and monosubstituted aromatic (phenyl). These are illustrated in the problems that follow.

Topic Test 4: Nuclear Magnetic Resonance and ¹H-NMR

True/False

1. The peaks in the proton NMR spectrum of bromoethane integrate in the ratio 2:3.

2. The peaks in the proton NMR spectrum of CH$_3$CH$_2$OCH$_2$CH$_3$ integrate in the ratio 2:3.

3. 2-Bromo-2-methylpropane shows only a singlet in its proton NMR spectrum.

4. The proton NMR spectrum of ethane appears as a quartet because each methyl group has three neighboring protons.

Multiple Choice

5. Which of the following can be examined directly by NMR?
   a. ^{12}C
   b. ^{16}O
   c. ^{4}He
   d. All of the above
   e. None of the above

6. Which isomer below would you expect to show a proton NMR signal near 7 ppm that integrates for five protons?
   a. p-Xylene (p-dimethylbenzene)
   b. m-Xylene
   c. o-Xylene
   d. Ethylbenzene
   e. All of the above

Short Answer

7-9. Assume you have an infinitely well-resolved proton NMR spectrum for each of the following. How many environments would you expect to see for each?

7.   8. CH₃CH₂—⌬—NO₂   9. CH₃—⌬—CH=CH₂ (with H,H,H shown)

10. Propose a structure for a compound that has the formula $C_4H_{10}O$ and shows the proton NMR spectrum tabulated below.

δ	Multiplicity	Relative Integral
3.55	Septet	1H
3.30	Singlet	3H
1.2	Doublet	6H

11. Propose a structure for a compound that has the formula $C_{10}H_{14}$ and shows the proton NMR spectrum tabulated below.

δ	Multiplicity	Relative Integral
7.1	Broad singlet	5H
1.2	Singlet	9H

Topic Test 4: Answers

1. **True.** There are two proton environments in CH_3CH_2Br. There are two of one type and three of the other.

2. **True.** There are two proton environments in $(CH_3CH_2)_2O$. There are four of one type and six of the other, but because the integrator can only express the ratios (not absolute numbers) this will appear as 2:3.

3. **True.** All the protons of $(CH_3)_3CBr$ are equivalent and are not coupled or split.

4. **False.** All six hydrogens of ethane are equivalent and do not couple to one another. The proton NMR spectrum of ethane will appear as a singlet.

5. **e.** The three nuclei mentioned have even masses and atomic numbers.

6. **d.** This is the only monosubstituted benzene listed. For the peak near 7 ppm (aromatic) to integrate for 5H, there can only be one substituent n the ring.

7. Four

8. Four

9. Eight

10. The 3H singlet is an isolated methyl. The 1H septet and 6H doublet are coupled to one another and are the characteristic pattern of an isopropyl group. The formula indicates saturation so the only reasonable structure is 2-methoxypropane: $(CH_3)_2CHOCH_3$.

11. The 9H singlet is characteristic of a *t*-butyl group and the 5H signal near 7 ppm indicates a monosubstituted benzene (phenyl). The formula indicates the degree of unsaturation is four. The two fragments account for the entire structure of *t*-butylbenzene: Ph—$C(CH_3)_3$.

TOPIC 5: CARBON-13 NMR

KEY POINTS

✓ *How is carbon-13 NMR like proton NMR?*

✓ *How is carbon-13 NMR different from proton NMR?*

✓ *How does one interpret a carbon-13 NMR spectrum?*

Carbon-13 NMR spectroscopy is similar to proton NMR in that the number of peaks in the spectrum normally corresponds to the number of different carbon environments and the chemical shifts of the carbon signals provide some indication of the nature of each environment. Carbon-13 NMR differs from proton NMR in that integration is normally not done and the magnitude of each resonance signal depends not only on the number of carbons that produced it but on several other factors that make integration unreliable. The chemical shift range of carbon-13 NMR signals is approximately 220 ppm and is expressed as δ (downfield of TMS). This wide range of shifts makes it less likely that accidental superimposition of peaks will occur. **Table 9.5** shows carbon-13 NMR chemical shift ranges.

Table 9.5 Some Approximate Carbon-13 NMR Chemical Shift Ranges (δ)	
C=O	180–220
Aromatic	120–170
Alkene C=C	100–150
Alkyne C≡C	70–90
C—O	30–85
C—N	10–60
C—Cl	40–85
C—Br	30–70
—CH$_2$— (alkane)	20–65
CH$_3$— (alkane)	10–40

Coupling between adjacent carbons is not generally observed. Carbon-13 NMR spectra are usually acquired in a mode that removes the effect of any coupling to the protons attached to the carbons. Spectra run in this **proton-decoupled** mode show each carbon environment as a singlet. It is possible to establish the number of protons on each carbon by running the spectrum in the coupled mode producing a spectrum in which each carbon signal will be slit into an n + 1 multiplet resulting from the attachment of n protons. The number of attached protons is also often determined by more modern NMR techniques.

Topic Test 5: Carbon-13 NMR

True/False

1. The carbon-13 NMR spectrum of 2-methylpropane has only two lines.
2. A proton-decoupled (normal) carbon-13 NMR spectrum shows splitting due to the coupling of adjacent carbons.

Multiple Choice

3. Which isomer below will show a carbon-13 NMR signal between 180 and 220 ppm?

   a. ▷—CH$_2$OH   b. (cyclobutanone/oxetane structure)   c. (acetone-like structure with C=O)

d. All of the above
e. None of the above

4. The number of protons attached to a given carbon
   a. is not revealed in a proton-decoupled carbon-13 NMR spectrum.
   b. is revealed if the carbon-13 NMR spectrum is run in the proton-coupled mode.
   c. can be detected using modern NMR techniques.
   d. All of the above
   e. None of the above

Short Answer

5–6. Identify the number of different carbon environments in the compounds below.

5. [structure: ortho-disubstituted benzene with two propyl-like chains]

6. [structure: carboxylic acid with alkene and branched alkyl chain]

Topic Test 5: Answers

1. **True:** $(CH_3)_3CH$.

2. **False.** There is generally no coupling between adjacent carbons because the natural abundance of carbon-13 is so low (about 1%) and therefore the statistical likelihood that two carbon-13 nuclei will be bounded to one another is small.

3. **c.** A signal between 180 and 220 ppm is likely a carbonyl. The four-carbon ketone is the only carbonyl-containing compound shown.

4. **d**

5. [labeled structure with carbons 1–6] six

6. [labeled structure with carbons 1–8] eight

APPLICATION

Organic chemists have been using NMR since the 1960s, but more recently a popular technology based on the NMR phenomenon has been directly applied to the field of medicine. Magnetic resonance imaging provides a noninvasive method for imaging tissues from deep within a patient's body. The patient is placed within the poles of a large magnet and a series of "spectra" are acquired corresponding to cross-sections of all or part of the body. The data are processed by computer and can be displayed as a two-dimensional image or even a three-dimensional virtual reality image.

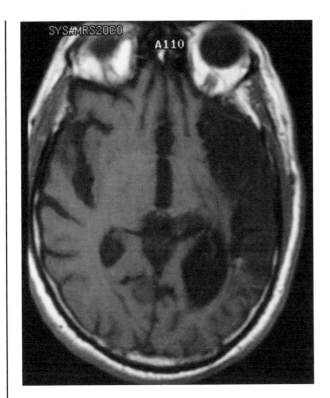

From Mills VM, Cassidy JW, Katz, DI. Neurologic rehabilitation; a guide to diagnosis, prognosis, and treatment planning. Malden, MA: Blackwell Scientific, 1997.

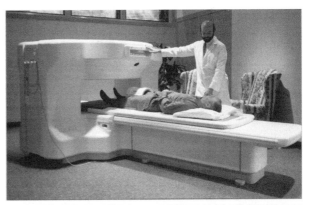

The two figures above are courtesy of the Shields Health Care Group, Brockton, Massachusetts.

DEMONSTRATION PROBLEM

Deduce a structural formula from spectral data provided.

MS:	$M^+ = 134$		
IR:	Strong peak at $1703\,cm^{-1}$		
^{1}H-NMR:	ppm	Multiplicity	Integral
	10.0	Singlet	1H
	7.9	Doublet	2H
	7.4	Doublet	2H
	2.7	Quartet	2H
	1.3	Triplet	3H
^{13}C-NMR:	15, 29, 128, 130, 135, 152, and 192 ppm		

Solution

One strategy for solving problems of this type is to summarize what you see easily and then assemble possible structures from the detected "pieces." The IR peak at $1703\,cm^{-1}$ is probably a carbonyl. The singlet at 10 ppm in the proton NMR agrees with the carbonyl above. There are two recognizable patterns in the proton NMR. The two doublets in the aromatic region that each integrate for 2H each (or a total of 4H) indicate a *para*-substituted benzene ring. The 2H triplet and the 3H quartet are the classic combination indicating an ethyl group. Subtracting the pieces from what we know to be the molecular weight will reveal if all the parts have been identified.

$$
\begin{array}{rl}
134 & \text{M}^+ \\
-29 & \text{Aldehyde, O=C-H} \\
\hline
105 & \\
-76 & \text{p-aromatic, C}_6\text{H}_4 \\
\hline
29 & \\
-29 & \text{ethyl, CH}_3\text{CH}_2 \\
\hline
0 &
\end{array}
$$

All the mass has been accounted for. The pieces can only attach together in one way, thus giving *p*-ethylbenzaldehyde.

$$\text{CH}_3\text{CH}_2-\text{C}_6\text{H}_4-\text{CHO}$$

Now confirm that a proposed structure agrees with all available data. The carbon-13 NMR data were not used to obtain the solution, but they do agree with the proposed structure. The symmetry of the compound leads to only seven carbon environments (even though there are a total of nine carbons). There is a carbonyl (192 ppm), four aromatic signals, and two nonaromatic signals.

Chapter Test

True/False

1. The tallest (most intense) peak in a mass spectrum indicates the compound's MW.
2. A compound that has λ_{max} at longer than 210 nm is probably conjugated.
3. 1,4-Cyclohexadiene has only two peaks in its proton-decoupled carbon-13 NMR spectrum.
4. Protons on aromatic rings normally have a chemical shift near 5 ppm.
5. The protons on C(2) of propane will appear as a septet in the proton NMR spectrum.
6. Methylcyclohexane shows five lines in the carbon-13 NMR spectrum, but cyclohexane shows only one.

Multiple Choice

7. A mass spectral peak at M-17 probably indicates
   a. loss of OH.
   b. loss of Br.
   c. loss of methyl.
   d. Any of the above
   e. None of the above

8. Which compound below has a λ_{max} most like that of 1,3-pentadiene?
   a. $(\text{CH}_3)_2\text{CHCH}_2\text{CH}_2\text{CH}_2\text{CH}=\text{CH}-\text{CH}=\text{CH}_2$
   b. $\text{CH}_2=\text{CHCH}_2\text{CH}=\text{CH}_2$

c. $CH_3CH_2CH_2CH=CH_2$
   d. $CH_3CH_2CH_2CH_2CH_3$
   e. All of the above have essentially the same λ_{max}.

Indicate the kind of spectroscopy best associated with each of the following:

9. Stretching and bending of bonds.

10. Electrons are excited to higher energy molecular orbitals.

11. Molecules are bombarded by electrons to yield ions that subsequently fragment.

12. Spinning nuclei exposed to an applied magnetic field are excited to a higher energy state by a radio frequency.

13. Usually reveals the molecular weight of a compound.

14. Indicates the presence of various functional groups.

15. Provides information on the number of different hydrogen or carbon environments.

16. Shows the presence and extent of conjugation.

Combination Spectral Problems

Deduce structural formulas from the spectral data provided.

17. Mass spectrum shows $M^+ = 72$, IR shows a strong peak near $1720\,cm^{-1}$. Carbon-13 NMR shows four lines. The proton NMR is tabulated below.

δ	Multiplicity	Relative Integral
2.4	Quartet	2H
2.1	Singlet	3H
1.1	Triplet	3H

18. $M^+ = 108$; IR 1500, 1600, $1250\,cm^{-1}$; ^{13}C-NMR shows five lines; λ_{max} near 220, 270, and 285 nm

δ	Multiplicity	Relative Integral
6.8–7.2	Multiplet	5H
3.7	Singlet	3H

19. $M^+ = 122$; IR peaks near 1500, 1600 strong broad peak near $3350\,cm^{-1}$; ^{13}C-NMR shows six lines

δ	Multiplicity	Relative Integral
7.15	Multiplet	5H
3.75	Triplet	2H
2.75	Triplet	2H

20. $M^+ = 114$; IR near 2960, 2980, and strong near $1715\,cm^{-1}$; ^{13}C-NMR shows three lines (one is > 200 ppm)

δ	Multiplicity	Relative Integral
3.7	Septet	1H
1.0	Doublet	6H

21. M⁺ = 117; $\lambda_{max} \cong 234$ nm; ^{13}C-NMR shows six lines; IR 2230 cm^{-1}

δ	Multiplicity	Relative Integral
7.5	Doublet	2H
7.2	Doublet	2H
2.4	Singlet	3H

Chapter Test: Answers

1. **F** 2. **T** 3. **T** 4. **F** 5. **T** 6. **T** 7. **a** 8. **a**
9. IR
10. UV-vis
11. MS
12. NMR
13. MS
14. IR
15. NMR
16. UV-vis

17. CH₃CH₂–C(=O)–CH₃

18. C₆H₅–OCH₃

19. C₆H₅–CH₂CH₂OH

20. (CH₃)₂CH–C(=O)–CH(CH₃)₂

21. CH₃–C₆H₄–C≡N

Check Your Performance

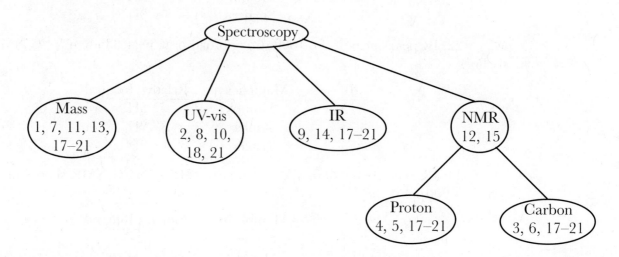

182   Chapter 9   Spectroscopy

Final Exam

Multiple Choice

1. Which reaction below is concerted?
   a. E2
   b. S_N2
   c. Diels-Alder
   d. All of the above
   e. None of the above

2. A strong peak between 1700 and 1760 cm^{-1} in an infrared spectrum most likely indicates
   a. the presence of an alkene.
   b. the presence of a saturated compound.
   c. the presence of an alkene.
   d. an aromatic structure.
   e. the presence of a carbonyl.

3. 1,4 Addition of one equivalent of HBr to 2,4-hexadiene yields
   a. 1-bromo-2-hexene.
   b. 1-bromo-3-hexene.
   c. 4-bromo-2-hexene.
   d. 2-bromo-3-hexene.
   e. None of the above

4. Which mechanism involves a carbocation intermediate?
   a. E2
   b. S_N2
   c. Nucleophilic aromatic substitution
   d. All of the above
   e. None of the above

5. Which compound has a degree of unsaturation = 5?
   a. *p*-Chlorophenol
   b. 2,4-Dinitroaniline
   c. *m*-Iodobenzaldehyde
   d. All of the above
   e. None of the above

6. Which compound shows the longest λ_{max} in the UV-vis spectrum?
   a. Toluene
   b. Benzoic acid
   c. *p*-Vinylbenzoic acid
   d. Styrene
   e. Benzene

7. Which compound would show two singlets for its ^{1}H-NMR spectrum?
   a. 1-Bromo-2,2-dimethylpropane
   b. 1,4-Dibromobutane
   c. 3,3-Dimethylpentane

d. All of the above
e. None of the above

8. Which of the following would undergo an E2 reaction with base the fastest?
   a. Ethylcyclohexane
   b. *cis*-1-Chloro-4-ethylcyclohexane
   c. *trans*-1-Chloro-4-ethylcyclohexane
   d. All the above will react by the E2 path at the same rate
   e. None of the above can react by the E2 pathway

9. Two diasteromers will
   a. always show the same mass peak for their mass spectral molecular ion, M^+.
   b. always have the same IR spectrum.
   c. always have the same ^{13}C-NMR spectrum.
   d. All of the above
   e. None of the above

10. The number of ^{13}C-NMR peaks one expects to see for *meta-tert*-butylbenzaldehyde is
    a. 8
    b. 9
    c. 10
    d. 11
    e. None of the above

11. How many resonance forms are there for the carbocation intermediate in the addition of HCl to styrene to yield 1-chloro-1-phenylethane?
    a. 2
    b. 3
    c. 4
    d. 5
    e. None of the above

12. Which reaction below is an example of electrophilic aromatic substitution?
    a. Bromobenzene reacted with Mg, ether
    b. Toluene reacted with excess H_2 over Rh catalyst
    c. *p*-Bromotoluene reacted with hot aqueous acid and $KMnO_4$
    d. Toluene reacted with Br_2, $FeBr_3$
    e. None of the above

13. When (*R*)-2-bromopentane is treated with HC≡C:⁻, there are products from both S_N2 and E2 present. The S_N2 product is
    a. (*R*)-3-Methyl-1-hexyne
    b. (*S*)-3-Methyl-1-hexyne
    c. 1-Pentene
    d. (*E*)-2-Pentene
    e. (*Z*)-2-Pentene
    f. None of the above

14. When 2-bromopentane is treated with HC≡C:⁻, there are products from both S_N2 and E2 present. The E2 product is
    a. (*R*)-3-Methyl-1-hexyne
    b. (*S*)-3-Methyl-1-hexyne

c. 1-Pentene
d. (*E*)-2-Pentene
e. (*Z*)-2-Pentene
f. None of the above

Provide an unambiguous structural formula for the major organic product(s) from reaction of 3-bromo-3-ethylpenatane with each of the following reagents and/or conditions.

15. $CH_3CH_2O^-K^+$, CH_3CH_2OH

16. Li, pentane then CuI, ether

17. Mg, ether then D_2O

18. $[(H_2C=CH)_2Cu]^-Li^+$

Provide an unambiguous structural formula for the major organic product(s) from reaction of $(CH_3CH_2)_2CHOH$ with each of the following reagents and/or conditions.

19. PBr_3, ether

20. Concentrated H_3PO_4, heat

21. $SOCl_2$

Provide an unambiguous structural formula for the major organic product(s) resulting from an EAS reaction of one equivalent of each of the following reagents with the compound shown below.

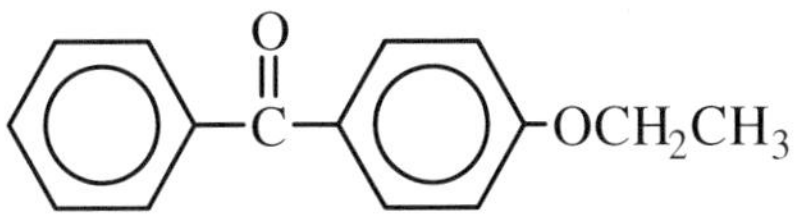

22. HNO_3, H_2SO_4

23. SO_3, H_2SO_4

Provide an unambiguous structural formula for the major organic product(s) resulting from an EAS reaction of one equivalent of each of the following reagents with the compound shown below.

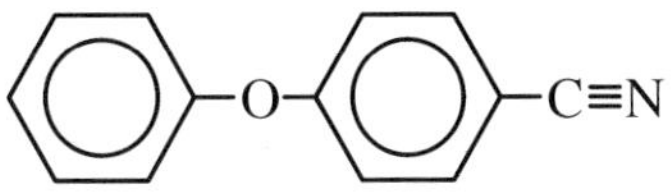

24. $CH_3CH_2C(=O)Cl$, $AlCl_3$

25. Br_2, $FeBr_3$

Provide an unambiguous structural formula for the major organic product(s) from reaction of cumene (isopropylbenzene) with each of the following reagents and/or conditions.

26. Excess H_2, Rh/C

27. *N*-Bromosuccinimide (NBS), peroxide or light

28. $KMnO_4$, H_3O^+, heat

Provide unambiguous structural formulas for the missing organic compounds.

29. [p-benzoquinone] + excess CH$_2$=CHCH=CH$_2$ →

30. ___ + ___ —[4+2]→ [7-oxabicyclic product with two CF$_3$ groups]

Final Exam Answers

1. **d** 2. **e** 3. **d** 4. **e** 5. **c** 6. **c** 7. **a** 8. **b** 9. **a** 10. **b** 11. **c** 12. **d**
13. **b** 14. **d**

15. ((CH$_3$CH$_2$)$_2$C=CHCH$_3$

16. [(CH$_3$CH$_2$)$_3$C]$_2$$\overset{\ominus}{\text{Cu}}$$\overset{\oplus}{\text{Li}}$

17. (CH$_3$CH$_2$)$_3$CH

18. (CH$_3$CH$_2$)$_3$CCH=CH$_2$

19. [sec-butyl bromide structure] Br

20. CH$_3$CH$_2$CH=CHCH$_3$ (*cis* and *trans*)

21. [sec-butyl chloride structure] Cl

22. [PhC(O)–C$_6$H$_3$(NO$_2$)–OCH$_2$CH$_3$]

23. [PhC(O)–C$_6$H$_3$(SO$_3$H)–OCH$_2$CH$_3$]

24. CH$_3$CH$_2$C(O)–C$_6$H$_4$–O–C$_6$H$_4$–C≡N
    (or *ortho* on same ring)

25. Br–C$_6$H$_4$–O–C$_6$H$_4$–C≡N
    (or *ortho* on same ring)

26. [isopropylcyclohexane]

27. C$_6$H$_5$–C(CH$_3$)$_2$–Br

28. C$_6$H$_5$–CO$_2$H

29. [octahydroanthracene-9,10-dione with two remaining alkene double bonds]

30. [furan] + CF$_3$–C≡C–CF$_3$

186   Final Exam

INDEX

Acetylide, 80
Achiral, 92
Acid-base reactions, 14
Acids, 14
Activating groups, 154–155
Acyl halide, 38
Acylation, aromatic, 153
Addition reactions
    1,2 addition, 136–137
    1,4 addition, 136–137
    to alkenes, 48–57
Alcohol, 38
    dehydration of, 64
    from alkenes, 51
    reaction with concentrated H_3PO_4, 64–65
    reaction with concentrated H_2SO_4, 64–65
    reaction with HX, 48–49, 60
    reaction with PBr_3, 116
    reaction with $POCl_3$, 64–65
    reaction with $SOCl_2$, 116
Aldehyde, 38
    chemical shift of, NMR, 174
    from alkenes, 58
    from alkynes, 76
    synthesis of, 58, 76
Alkanes, 21–35
    boiling points of, 29
    bromination of, 34–35
    chlorination of, 34–35
    combustion of, 35
    conformations of, 30–32
    defined, 21
    from alkenes, 55–56
    from alkynes, 78
    from Grignard reagents, 127
    halogenation of, 34–35
    IR of, 171
    isomers, 24, 89–90, 94
    melting points of, 29
    nomenclature of, 21–24, 26–27
    properties of, 29
    reaction with Br_2, 34–35
    reaction with Cl_2, 34–35
    representations of, 24–25
    structure of,
    unbranched, 21–22, 29
    van der Waals forces in, 29
Alkenes, 38
    addition reactions of, 48–57
    alcohols from, 51
    alkanes from, 55–56
    alkyl bromide from, 48–49
    allylic bromination, 116
    bonding orbitals in, 43
    bromohydrins from, 53–54
    bromonium ions from, 53–54
    cis-trans isomers of, 44
    cleavage of, 57–58
    1,2 diols from, 58
    electrophilic addition reactions of, 48–57
    from alcohols, 64–65
    from alkyl halides
    from alkynes
    halohydrins from, 53–54
    hydration of, 51
    hydroboration of, 51
    hydrogenation of, 55–56
    IR spectroscopy of, 171
    isomers, 44
    ketones from, 57–58
    Markovnikov's rule and, 48
    molecular orbitals of, 43
    nomenclature of, 44
    oxymercuration of, 51
    ozonolysis of, 58
    polymerization of, 62
    radical addition of HBr to, 60
    reaction with borane, 51
    reaction with Br_2, 53–54
    reaction with Cl_2, 53–54
    reaction with HBr, 48–49, 60
    reaction with $KMnO_4$, 57–58
    reaction with mercuric acetate, 51
    reaction with NBS, 116
    reaction with OsO_4, 58
    reaction with ozone, 58
    reduction of, 55–56
    stereochemistry of addition to, 53–54
    structure of, 43
    trialkylboranes from, 51
Alkene cleavage reactions, 57–58
Alkyl anion, 80
Alkyl bromide, 48–49, 60
    from alkenes, 48–49, 60
    synthesis of, 48–49, 60, 116
Alkyl chloride, synthesis of, 116
Alkyl groups, 22
    effect on EAS reaction, 154–155
    nomenclature of, 22
Alkyl halides, 113–131
    alkenes from, 64–65, 120–122
    alkynes from, 72
    common names of, 113–114
    elimination reactions of, 64–65, 72, 120–122
    from alcohols, 116
    from alkenes, 115–116
    Gilman reagents from, 127
    Grignard reagents from, 125–126
    nomenclature, 113–114
    organolithium reagents from, 126
    preparation of, 115–116
    reaction with acetylide anions, 80
    reaction with Gilman reagents, 126
    reaction with KOH, 64–65, 72, 120–122
    reaction with lithium, 126

reaction with magnesium, 125–126
structure of, 113–114
synthesis of, 115–116
Alkyl shift in carbocations, 49
Alkylation
of acetylides, 80
of aromatic compounds, 153
Alkylbenzenes, 159
benzylic halogenation of, 159
oxidation of, 159
reaction with KMnO$_4$, 159
reaction with NBS, 159
Alkylboranes, 51
Alkyllithium reagents, 126
Alkynes
acidity of terminal, 80
aldehydes from, 76
alkanes from, 78
carboxylic acids from, 82
cis alkenes from, 78
cleavage of, 82
from dihalides, 72
hydration of, 76
hydroboration of, 76
hydrogenation of, 78
IR spectroscopy of, 171
ketones from, 76
nomenclature of, 72
reaction of terminal with base, 80
reaction with Br$_2$, 74
reaction with H$_2$, 78
reaction with HX, 74
reaction with KMnO$_4$, 82
reduction of, 78
synthesis of, 72, 80
trans alkenes from, 78
Allyl group, 45
Allylic, 116
Allylic bromination, 116
Allylic carbocations, 137
Amide, 38
Amine, 38
Angle strain, 31
Anhydride, 38
Aniline, 151
Anti conformation, 31

Anti-periplanar, 121
Aromatic compounds, 148
alkylation of, 153
bromination of, 153
characteristics of,
chlorination of, 153
common names of, 151
electrophilic substitution of, 153
Friedel-Crafts reactions of, 153
IR spectroscopy of, 171
nitration of, 153
NMR spectroscopy of, 174, 177
nomenclature of, 151
ortho, meta, para prefixes for, 151
oxidation of side chains, 159
reduction of, 159
side chain reactions of, 159
sulfonation of, 153
Aromatic hydrogen, chemical shift of, 174
Arrow, electron flow of, mechanism, 48
Axial position, 31–32

Base peak of mass spectrum, 168
Bases, 14
Benzaldehyde, 151
Benzene, 148–149
alkylation of, 153
Friedel-Crafts reaction of, 153
Kekule structure for, 149
nitration of, 153
reaction with Br$_2$, 153
reaction with Cl$_2$, 153
reaction with HNO$_3$, 153
reaction with SO$_3$, 153
reduction of, 159
resonance in, 148–149
stability of, 148–149
sulfonation of, 153
Benzenesulfonic acid, 151
Benzoic acid, 151
Benzylic position, 159
Benzyne, 157
synthesis of, 157
Boat, cyclohexane, 31
Boiling points of alkanes, 29
Bond angle, 8

Bonds
covalent, 4
ionic, 4
polar covalent, 4
pi, 6
polarity, 15
sigma, 6
Borane
reaction with alkenes, 51
reaction with alkynes, 76
Bromine, reaction with
alkanes, 34–35
alkenes, 53–54
alkynes, 74
aromatic compounds, 153
Bromohydrin, 53–54
Bromonium ion, 53–54
Bronsted acids and bases, 14

Carbocation, 49
alkyl shifts in, 49
Friedel-Crafts reaction of, 153
hydride shifts in, 49
rearrangements, 49
stability, 49
Carbonyl group
^{13}C NMR of, 177
IR spectroscopy of, 171
Carboxylic acid, 38
Chair, cyclohexane, 31
Chemical shift, NMR, 173–174, 177
Chiral, 92
Chirality, 92
Chlordane, 128
CIP (Cahn, Ingold, Prelog) priorities, 44, 96
Cis, 27, 44
Combustion, 35
Configuration, *S* and *R*, 96–97
Conformation, 30, 139–140
anti, 31
butane, 31
cyclohexane, 31
gauche, 31
syn, 31
Conformational isomers, 30
anti, 31

butane, 31
cyclohexane, 31
eclipsed, 31
gauche, 31
syn, 31
Conformers, 30–31
anti, 31
boat, 31
butane, 31
chair, 31
cyclohexane, 31
eclipsed, 31
gauche, 31
syn, 31
twist boat, 31
Conjugate acid, 14
Conjugate base, 14
Conjugated diene, 133–134
1,4 addition to, 136–137
allylic carbocation from, 137
Diels-Alder reaction of, 139
reaction with HBr, 136–137
stability of, 134
Conjugation, 133–134
Constitutional isomers, 89–90
Coupling in NMR spectra, 174
Covalent bond, 4
Cycloaddition, 139
Cycloalkanes, 26–27
Cyclohexane conformations, 31
Cyclopentadienyl anion aromaticity, 149

DDT, 128
Deactivating groups, 154–155
Degree of unsaturation, 46–47
Dehydration, 64–65
of alcohol, 64–65
Zaitsev's rule for, 64
Dehydrohalogenation, 64–65, 72
Dextrorotatory, 101
Diastereomers, 94
Diels-Alder reactions, 139–140
Dienes, conjugated, 133–134
1,2 vs. 1,4 addition to, 136–137
Diels-Alder reactions of, 139–140

Dienophile, 139–140
Dihedral angle, 31
Diols
cleavage of, 58
from alkenes, 58
reaction with HIO_4, 58
Diorganocopper reagents. *See* Gilman reagents
Dipoles, 15–16
Downfield, NMR, 73

E and Z isomers, 44
E1, 121–122
E2, 121
Eclipsed conformation, 31
Electron donating groups, 154–155
Electron dot structure, 1–2
Electron withdrawing groups, 154–155
Electronegativity, 4
Electrophile, 48, 153
Electrophilic addition, 48–54
Electrophilic aromatic substitution, 153
mechanism, 153
substituent effects in, 154–155
Elimination reactions, 64–65, 72, 120–122
Enantiomers, 94
Enol, 76
Equatorial, 31–32
Ester, 38
Ethene, structure, 43
Ether, 38

Fingerprint region, IR, 171, 172
Fischer projections, 98–99
Formal charge, 10
Fragmentation in mass spectrometry, 167–168
Free-radical halogenation of alkanes, 34–35
Friedel-Crafts acylation, 153
Friedel-Crafts alkylation, 153
Functional groups, 37–38
IR spectroscopy of, 171
table of, 38

gauche conformation, 31
Gasoline octane rating, 39
Gilman reagents, 126
coupling reactions of, 126
from alkyl halides, 126
reaction with alkyl halides, 126
Grignard reagents, 125–126
alkanes from, 126
from alkyl halides, 125–126
polarity of, 126
reaction with acid, 126

Halogenation of alkenes, 53
Halohydrin, 53–54
Halonium ion, 53
Hückel number, 148
Hybrid orbitals, 7–8
Hybridization, 7–8
Hydration of
alkenes, 51
alkynes, 76
Hydroboration
of alkenes, 51
of alkynes, 76
regiochemistry of, 51, 76
Hydrocarbon, defined, 21
Hydrogen peroxide, reaction with alkylboranes, 51, 76
Hydrogenation of
alkenes, 56
alkynes, 78

Ibuprofen, 105–106
Infrared spectroscopy, 171
Insecticides, 128
Integration, NMR, 174
Inversion, 118
Ionic bonding, 4
Isomers
cis and *trans*, 27, 44, 94
constitutional, 89–90
defined, 24
E and Z, 44
R and S, 96–97
stereoisomers, 90
Isoprene, 142
Isoprene rule, 142

Isopropyl group, 45
IUPAC, 23–24

Keto-enol tautomerism, 76
Ketone, 38
Kinetic control, 136–137

Leaving group, 118
Levorotatory, 101
Lewis acids and bases, 14
Lewis dot structures, 1
Lindane, 128
Lindlar's catalyst, 78
Line-bond structures, 24–25
Lithium diorganocopper reagents. *See* Gilman reagents

Magnetic resonance imaging (MRI), 178–179
Markovnikov's rule, 48
Mass spectrometry, 167–168
Mechanisms
   1,2 addition, 136–137
   1,4 addition, 136–137
   alkene halogenation with X_2, 53
   alkene + HX, 38–49
   alkene hydration, 51
   alkene hydroboration, 51
   benzyne, 157
   bromohydrin formation, 54
   conjugate addition, 136–137
   Diels-Alder, 139–140
   E1, 121–122
   E2, 121
   electrophilic addition, 48–54
   electrophilic aromatic substitution, 153
   halohydrin formation, 53–54
   HBr radical addition to alkenes, 60
   hydroboration of alkenes, 51
   nucleophilic aromatic substitution, 157
   oxymercuration, 51
   radical halogenation of alkanes, 34–35
   S_N1, 119
   S_N2, 118
   stereochemistry of, 104
Melting points, alkanes, 29
meso, 92
meta, 151
Meta directors, 154
Mirror image, 90, 92
Molecular ion, mass spectrometry, 167
Monomers, 62
MRI, 178–179
Multiplets, NMR, 174

n + 1 splitting in NMR, 174
NBS (*N*-bromosuccinimide), 116, 159
Newman projections, 31
Nitration, aromatic, 153
Nitrile, 38
   IR spectroscopy of, 171
NMR, 173–174, 176–177
   carbon, 176–177
   proton, 173–174
Nomenclature
   alkanes, 22, 24
   alkenes, 44
   alkyl groups, 22
   alkyl halides, 113–114
   alkynes, 72
   aromatics, 150–152
   cycloalkanes, 26
Nucleophile, 48

Octane rating, gasoline, 39
Octet rule, 1
Optical activity, 101
Orbitals, 5, 133–134
   atomic, 5–6
   hybrid, 7–8
   overlap in conjugated systems, 133–134
Organic chemistry defined, 1
Organolithium compounds, 126
Organometallics, 125–126
ortho, 151
ortho/para directors, 154
Osmium tetroxide, reaction with alkenes, 58
Oxidation defined, 55–56
Oxymercuration of alkenes, 51
Ozonolysis
   alkenes, 58
   alkynes, 82

para, 151
Partial charge, polar covalent bonds, 15–16
Periodic acid, 58
Periplanar, 121
Phenol, 151
Phenyl group, 151–152
Phosphorous tribromide, 116
Plane of symmetry, 92
Plane-polarized light, 101
Polar covalent bond, 4
Polarimeter, 101–102
Polarimetry, 101–102
Polymerization of alkenes, 62
Polymers, 62, 67
Potassium permanganate
   reaction with alkenes, 57–58
   reaction with alkylbenzenes, 159
   reaction with alkynes, 82
Primary carbon, 28
Pyridine, aromaticity of, 149–150

Quaternary carbon, 28

R and S configuration, 96–97
R as an alkyl group symbol, 37
Racemate, 101
   resolution of, 103–104
Racemic mixture, 101
   resolution of, 103–104
Radical addition of HBr to alkenes, 60
Radical cation, mass spectrometry, 167
Radical halogenation of alkanes, 34–35
Rate-limiting step, 119
Rearrangement, carbocations, 49
Reduction, defined, 55–56
   of alkenes, 55–56
   of alkynes, 78
   of aromatic compounds, 159

Resolution of a racemic mixture, 103–104
Resonance, 12, 149, 153
Resonance forms, 12, 149, 153
Ring flip, cyclohexane, 31
Rotation, specific, 101–102

S and R configurations, 96–97
Saturated, 21, 46
s-cis conformation, 139–140
Secondary carbon, 28
Shapes, molecular, 8
Skeletal structures, 24
S_N1, 119
S_N2, 118
Solvolysis, 119
sp hybrid orbitals, 8
sp^2 hybrid orbitals, 8
sp^3 hybrid orbitals, 8
Specific rotation, 101–102
Spectroscopy, 167–179
    infrared, 171
        mass, 167–168
            NMR, 173–174, 176–177
                carbon, 176–177
                proton, 173–174
    ultraviolet-visible, 169
Stereochemistry, 89–106
    of Diels-Alder reaction, 139–140
    of E1, 121–123
    of E2, 121–123
    of electrophilic addition reactions, 104
    of S_N1, 119
    of S_N2, 118
Stereogenic centers, 92
Stereoisomers, 90
    alkene, 44, 90
    calculating maximum number of, 95
    kinds of, 94
    properties of, 103
    substituted cycloalkanes, 27, 90
Strain, 31
s-trans diene conformation, 139–140
Substituent effects on EAS reactions, 154–155
Substrate, 118
syn-periplanar, 121

Tautomer, 76
Tautomerism, 76
Teflon, 67
Terpenes, 142
Terpenoids, 142
tert-butyl group, 22
Tertiary carbon, 28
Tetrahedral geometry, 8
Tetramethylsilane, 173
Thermodynamic control, 136–137
TMS, 173
TNT, 161
Toluene, 151
Torsional strain, 31
trans, 27
Twist boat, cyclohexane conformer, 31

Ultraviolet-visible spectroscopy, 169
Unsaturation, degree of, 46–47
Upfield, NMR, 73

Valence shell, 1–2
Van der Waals forces, 29
Vicinal, 53, 58
Vicinal dihalides, 53
Vicinal diols, 58
Vinyl group, 45
Vinylic halide, 126

Z and *E* isomers, 44
Zaitsev's rule, 64